P9-DFS-412

INTERACTIVE STUDY GUIDE

A COMPANION TO

CCSS Your Common Core Edition

GLENCOE MATH

ACCELERATED

A PRE-ALGEBRA PROGRAM

McGraw Hill Education

Bothell, WA • Chicago, IL • Columbus, OH • New York, NY

connectED.mcgraw-hill.com

The *McGraw·Hill* Companies

Education

Copyright © 2014 by The McGraw-Hill Companies, Inc.

STEM McGraw-Hill is committed to providing instructional
materials in Science, Technology, Engineering, and Mathematics
(STEM) that give all students a solid foundation, one that prepares
them for college and careers in the 21st century.
Send all inquiries to:

McGraw-Hill Education
8787 Orion Place
Columbus, OH 43240

ISBN: 978-0-07-664448-3
MHID: 0-07-664448-0

Printed in the United States of America.

13 14 15 16 LMN 22 21 20 19

Our mission is to provide educational resources that
enable students to become the problem solvers of
the 21st century and inspire them to explore careers
within Science, Technology, Engineering, and
Mathematics (STEM) related fields.

Chapter 4
Powers and Roots

Chapter 5
Ratio, Proportion, and Similar Figures

Chapter 6
Percents

UNIT 3
Introduction to Sampling and Inference

UNIT 4
Creating, Comparing, and Analyzing Geometric Figures

FOLDABLES® Study Organizer

What Are Foldables and How Do I Create Them?

Foldables are three-dimensional graphic organizers that help you create study guides for each chapter in your book.

Step 1 Find the Foldable for the chapter you are currently studying. Follow the cutting and assembly instructions at the top of the page.

Step 2 Go to the Key Concept Check at the end of the chapter you are currently studying. Match up the tabs and attach your Foldable to this page. Dotted tabs show where to place your Foldable. Striped tabs indicate where to tape the Foldable.

Step 1

Step 2

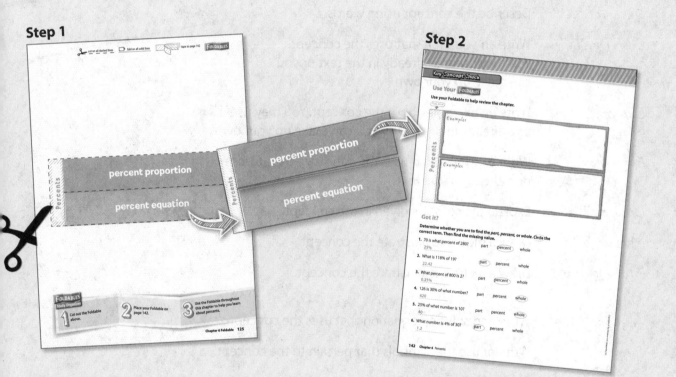

How Will I Know When to Use My Foldable?

When it's time to work on your Foldable, you will see a Foldables logo at the bottom of the **Rate Yourself!** box on the Notes pages. This lets you know that it is time to update it with concepts from that lesson. Once you've completed your Foldable, use it to study for the chapter test.

Rate Yourself!

How well do you understand the percent proportion? Circle the image that applies.

Clear   Somewhat Clear   Not So Clear

For more help, go online to access a Personal Tutor.

FOLDABLES *Time to update your Foldable!*

How Do I Complete My Foldable?

No two Foldables in your book will look alike. However, some will ask you to fill in similar information. Below are some of the instructions you'll see as you complete your Foldable. **HAVE FUN** learning math using Foldables!

Instructions and what they mean

Best Used to...	Complete the sentence explaining when the concept should be used.
Definition	Write a definition in your own words.
Description	Describe the concept using words.
Equation	Write an equation that uses the concept. You may use one already in the text or you can make up your own.
Example	Write an example about the concept. You may use one already in the text or you can make up your own.
Formulas	Write a formula that uses the concept. You may use one already in the text.
How do I ...?	Explain the steps involved in the concept.
Models	Draw a model to illustrate the concept.
Picture	Draw a picture to illustrate the concept.
Solve Algebraically	Write and solve an equation that uses the concept.
Symbols	Write or use the symbols that pertain to the concept.
Write About It	Write a definition or description in your own words.
Words	Write the words that pertain to the concept.

Meet Foldables Author Dinah Zike

Dinah Zike is known for designing hands-on manipulatives that are used nationally and internationally by teachers and parents. Dinah is an explosion of energy and ideas. Her excitement and joy for learning inspires everyone she touches.

Chapter 1
The Language of Algebra

Chapter Preview

Vocabulary

algebra	equation	numerical expression	simplify
algebraic expression	evaluate	order of operations	variable
coordinate plane	four-step plan	ordered pair	work backward
coordinate system	graph	origin	x-axis
counterexample	guess, check, and revise	properties	x-coordinate
deductive reasoning		range	y-axis
defining a variable	look for a pattern	relation	y-coordinate
domain	make a table		

Key Concept Activity

Look through the chapter. Write one or two key concepts from each lesson. Later, you can use this as a study tool for your chapter review.

Lesson	Key Concept(s)
1-1 A Plan for Problem Solving	
1-2 Words and Expressions	
1-3 Variables and Expressions	
1-4 Properties of Numbers	
1-5 Problem-Solving Strategies	
1-6 Ordered Pairs and Relations	
1-7 Words, Equations, Tables, and Graphs	

Are You Ready?

Try the Quick Check below.
Or, take the Online Readiness Quiz.

Check ✓

Quick Review

CCSS

Common Core Review 6.NS.3, 6.NS.6

Example 1

Find 11.9 − 2.15.

$$
\begin{array}{r}
\overset{810}{11.9\!0} \\
-\ 2.15 \\
\hline
9.75
\end{array}
$$

Annex a zero to align
the decimal points.

Subtract.

$11.9 - 2.15 = 9.75$

Example 2

Write the number that represents point A on the number line.

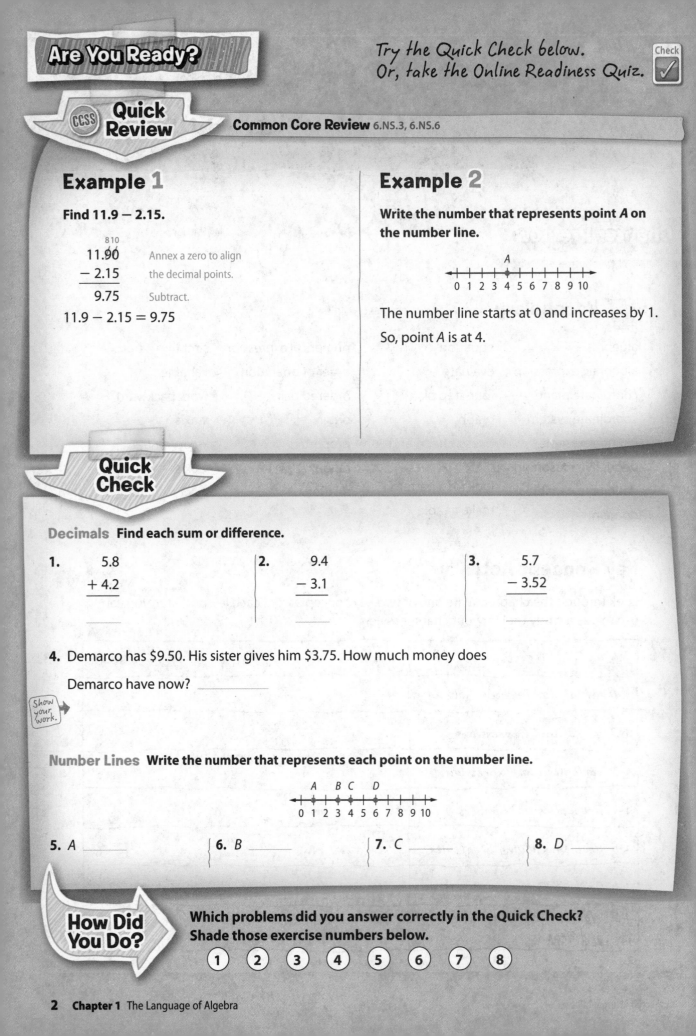

The number line starts at 0 and increases by 1.

So, point A is at 4.

Quick Check

Decimals Find each sum or difference.

1.
$$
\begin{array}{r}
5.8 \\
+\ 4.2 \\
\hline
\end{array}
$$

2.
$$
\begin{array}{r}
9.4 \\
-\ 3.1 \\
\hline
\end{array}
$$

3.
$$
\begin{array}{r}
5.7 \\
-\ 3.52 \\
\hline
\end{array}
$$

4. Demarco has $9.50. His sister gives him $3.75. How much money does

Demarco have now? _____

Show your work.

Number Lines Write the number that represents each point on the number line.

5. A _____ 6. B _____ 7. C _____ 8. D _____

How Did You Do?

Which problems did you answer correctly in the Quick Check?
Shade those exercise numbers below.

① ② ③ ④ ⑤ ⑥ ⑦ ⑧

✂ cut on all dashed lines   ▭ fold on all solid lines   tape to page 22   **FOLDABLES**

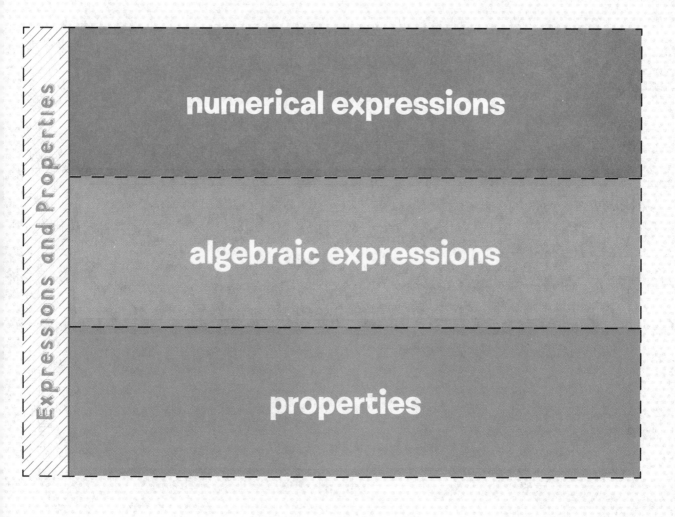

Expressions and Properties

numerical expressions

algebraic expressions

properties

FOLDABLES
Study Organizer

 1 Cut out the Foldable above.

2 Place your Foldable on page 22.

 3 Use the Foldable throughout this chapter to help you learn about expressions and properties.

Example

Example

Example

page 22

A Plan for Problem Solving

Getting Started

Scan Lesson 1-1 in your textbook. List two real-world scenarios in which you could use the four-step plan to solve a problem.

- _____
  _____

- _____
  _____

Quick Review

Circle the operation that you would use to answer each question.

- How many more?
  × −

- How many times as many? ÷ +

Real-World Link

Cell Phones The table shows the results of a survey about how teens use their cell phones. Fifty of the teens surveyed said they use their cell phones to play games *and* play music. How many play music only?

I Use My Cell Phone to...	Number of Teens
access social network sites	46
exchange videos	61
play games	92
play music	120
take pictures	177

1. Fill in the information that you know.

   [ ] teens use their cell phones to play games.

   [ ] teens use their cell phones to play music.

   [ ] teens use their cell phones to both play games and play music.

2. What operation(s) could you use to complete the diagram at the right? Explain.

   _____

   _____

   Play Games   Play Music

   [ ] 50 [ ]

3. Use the diagram to find the number of teens surveyed who use their cell phones to play music

   only. _____

4. Does your answer make sense? Explain. _____

   _____

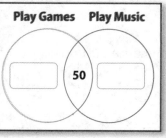

Notes

The Four-Step Plan

Identify the step in the four-step plan to which each statement or question belongs. Write U for *Understand*, P for *Plan*, S for *Solve*, or C for *Check*.

_____ Determine how the facts relate to each other.

_____ What facts do I know?

_____ Use my plan to solve the problem.

_____ What do I need to find out?

_____ Does my answer make sense?

_____ Choose one or more problem-solving strategies.

Solving Multi-Step Problems

Choose one of the multi-step problems in Lesson 1-1.

Describe your plan to solve the problem.

_____

_____

How many operations must be used? List them.

_____

_____

_____

Summary

Write 2–3 sentences to summarize the lesson.

_____

_____

_____

_____

_____

_____

Rate Yourself!

☐ I understand how to use the four-step plan.

▶▶ Great! You're ready to move on!

☐ I still have questions about using the four-step plan.

 No Problem! Go online to access a Personal Tutor.

Words and Expressions

Getting Started

Write the math and the real-world definitions of operation.

- math definition _____

  _____

- real-world definition _____

  _____

Quick Review

Name a mathematical operation and its inverse.

_____

_____

Vocabulary Start-Up

A **numerical expression** contains a combination of numbers and operations, such as addition, subtraction, multiplication, and division.

Complete the graphic organizer.

numerical expression

Describe it

Examples

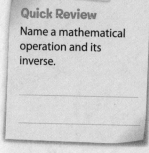

Real-World Link

Inventions Order the following steps for taking a digital picture from 1-5. Place a 1 by the step you would do first and a 5 by the step you would do last.

_____ Point the camera at the person or object to be photographed.

_____ Check the battery level.

_____ Play back your picture to see if you want to keep it.

_____ Take the picture.

_____ Zoom in or out to focus.

Notes

Translate Verbal Phrases into Expressions

Fill in each box with the correct symbol.

1. the difference of 18 and 13 ·····▶ 18 ☐ 13

2. the quotient of 81 and 9 ·····▶ 81 ☐ 9

3. the cost of 4 pencils at $0.30 each ·····▶ 4 ☐ 0.30

4. total students if there are 7 boys and 11 girls ·····▶ 7 ☐ 11

Order of Operations

Complete each step to evaluate 2[(7 + 9) × 3] − 15.

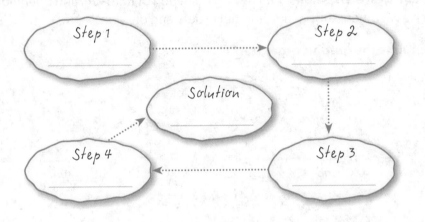

Summary

Write 2–3 sentences to summarize the lesson.

Rate Yourself!

How confident are you about writing and evaluating numerical expressions? Shade the ring on the target.

For more help, go online to access a Personal Tutor.

FOLDABLES Time to update your Foldable!

Variables and Expressions

Getting Started

Scan Lesson 1-3 in your textbook. List two headings you would use to make an outline of the lesson.

- _____
  _____

- _____
  _____

Vocabulary

Check the box next to the term or phrase below that you already know.

- ☐ algebra
- ☐ algebraic expression
- ☐ defining a variable
- ☐ Substitution Property of Equality

Real-World Link

Online Games There are several different types of online games, including role playing games, real-time strategy games, and social games. Morgan likes to play a certain social game with his friends. The table below shows the number of points earned for different actions in the game.

Online Gaming	
Action	**Points Earned**
jumping over an obstacle	6
feeding an animal	3
recruiting a new player	20

1. How can you find the total number of points Morgan would earn if he jumps over five obstacles and recruits two new players?

   _____

2. Refer to your answer for Exercise 1. How many points would he earn?

   _____

3. Let *a* represent *any number of animals*. What expression could be used to find the total points earned for jumping over one obstacle and feeding any number of animals? _____

4. Suppose the letter *p* is used to represent the number of players Morgan recruits to play the game. What would the expression $20p + 3a$ represent?

   _____

   _____

Notes

Algebraic Expressions and Verbal Phrases

Complete the graphic organizer to describe the steps involved in writing algebraic expressions.

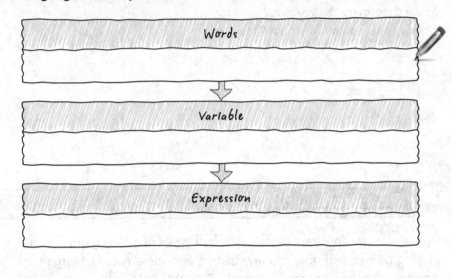

Words

Variable

Expression

Substitution Property of Equality

Evaluate each expression if $a = 3$, $b = 7$, and $c = 5$.

1. $6c \div 15 = \boxed{}$

2. $32 + 4a = \boxed{}$

3. $27a - (16 - 3c) = \boxed{}$

4. $\dfrac{bc}{a+2} = \boxed{}$

5. $2b - 4a = \boxed{}$

Summary

Write 2–3 sentences to summarize the lesson.

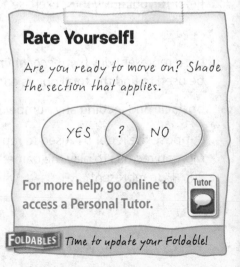

Rate Yourself!

Are you ready to move on? Shade the section that applies.

YES   ?   NO

For more help, go online to access a Personal Tutor.

FOLDABLES Time to update your Foldable!

Lesson 1-4

Properties of Numbers

Getting Started

Scan Lesson 1-4 in your textbook. Write the definitions of Associative Property of Addition and Associative Property of Multiplication in your own words.

- _____
  _____

- _____
  _____

Real-World Link

Crafts Did you know that duct tape comes in dozens of colors and in patterns ranging from tie-dye to digital camouflage? The table shows the amount of duct tape that is needed to make a purse.

Purse Panels	Length of Strips (in.)	Number of Strips
front and back	12	28
sides	4	14
bottom	14	4

1. Complete the steps below to find the total number of duct tape strips needed to make a purse.

   Method 1

   $(28 + 14) + 4$

   ⬚ $+ 4$

   ⬚

   Method 2

   $28 + (14 + 4)$

   $28 +$ ⬚

   ⬚

2. Based on what you learned from scanning the lesson, which property is illustrated in Exercise 1? _____

3. Write an expression to find the total length of duct tape needed to make each of the following. Then find each length.

   expression for side panels: _____   total length: _____

   expression for bottom panel: _____   total length: _____

4. What do you notice about the expressions and solutions that you wrote in Exercise 3? _____

Notes

Properties of Addition and Multiplication

(Circle) the word or number that describes each property. Then write an example for each.

Commutative (+ and ×)
order   grouping

Associative (+ and ×)
order   grouping

Identity (+)
0   1

Identity (×)
0   1

Simplify Algebraic Expressions

Simplify each expression by filling in the boxes with a variable or number.

1. $(8 + x) + 2$

☐ + ☐

2. $k \times (4 \times 4)$

☐ × ☐

3. $3 \times (7 \times p)$

☐ × ☐

4. $9 + (b + 5)$

☐ + ☐

Summary

Write 2–3 sentences to summarize the lesson.

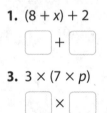

Rate Yourself!

How confident are you about identifying and using properties? Check the box that applies.

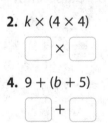

For more help, go online to access a Personal Tutor.

Tutor

FOLDABLES *Time to update your Foldable!*

Mid-Chapter Check

Vocabulary Check

1. **CCSS** **Be Precise** Define *four-step plan*. List the steps in the plan. (Lesson 1)

_____

_____

Fill in the blank.

2. To evaluate an expression using the order of operations, first simplify

_____. (Lesson 2)

Skills Check and Problem Solving

3. Carmina is saving for a tablet computer that costs $595. She saves $25 each week. In how many weeks will she have enough money to buy the tablet? Use

the four-step plan to solve. (Lesson 1) _____

Evaluate. (Lesson 2)

4. $\dfrac{40 - 8}{5 + 3} =$ _____

5. $2(8 + 3) - 4 \times 5 =$ _____

Translate each phrase into an algebraic expression. (Lesson 3)

6. five less than a number

_____

7. twice as many students

_____

8. the number of tires on c cars

_____

9. **Standardized Test Practice** Which of the following is an example of the Commutative Property? (Lesson 4)

Ⓐ $18 + (10 + 5) = (18 + 10) + 5$

Ⓑ $18 + 10 + 5 = 18 + (10 + 5)$

Ⓒ $(4 \cdot 2) \cdot 10 = (2 \cdot 4) \cdot 10$

Ⓓ $(4 \cdot 2) \cdot 10 = 4 \cdot (2 \cdot 10)$

Tag, You're It!

The *fork length* of a shark is the length from the tip of the snout to the fork of the tail. Use the information on the note cards to solve each problem.

1. Write an expression to represent the total length of a hammerhead shark that has a fork length of f feet. _____

2. Evaluate the expression from Exercise 1 to find the total length of a hammerhead shark that has a fork length of 11.6 feet. _____

3. Write an expression to represent the average fork length of a tiger shark, given the average fork length s of a sandbar shark. _____

4. Evaluate the expression from Exercise 3 to find the average fork length of a tiger shark if the average fork length of a sandbar shark is 129 centimeters. _____

5. Write an expression to find the average fork length of a white shark with a total length of t centimeters. _____

6. The total length of a white shark is 204 centimeters. Evaluate the expression in Exercise 5 to find the approximate fork length of the white shark. _____

Tiger Shark
A study found that the average fork length of a tiger shark is 55 centimeters less than twice the average fork length of a sandbar shark.

Hammerhead Shark
The total length of a hammerhead shark is about 1.3 times the fork length.

White Shark
The fork length of a white shark is about 5.74 centimeters less than 0.94 times the total length t.

Career Project

It's time to update your career portfolio! Describe the skills that would be necessary for a shark scientist to possess. Determine whether this type of career would be a good fit for you.

_____

_____

_____

_____

_____

List several challenges associated with this career.

• _____

• _____

• _____

• _____

• _____

Lesson 1-5

Problem-Solving Strategies

Getting Started

Scan Lesson 1-5 in your textbook. List two headings you would use to make an outline of the lesson.

- _____
  _____

- _____
  _____

Real-World Link

Movie Snacks Natalie and her friends are at the concession counter at the movie theater. After buying a ticket, Natalie has $10 left to buy snacks. The prices of the items she is considering are shown below.

Popcorn $3.75 SODA $2.50 Nachos $4.50 Choco Candy $1.75 $2.50

1. Natalie wants to buy a box of popcorn, a soda, and a hot dog. How much money will she have left? _____

2. How could you find all of the ways that she could receive the correct change? _____

3. Complete the table to show all of the possible combinations for her change if she always receives a one dollar bill and no pennies.

Dollar	Quarter	Dime	Nickel
1	1	0	0

4. In how many different ways can Natalie receive her change? _____

Notes

Use Problem-Solving Strategies

Complete the table by writing a summary of each problem-solving strategy.

Strategy	Summary
guess, check, and revise	
look for a pattern	
make a table	
work backward	

Select an appropriate strategy that could be used to solve each problem.

1. Marcy can run one lap in 65 seconds. Each additional lap takes her 2 seconds longer to run than the previous lap. How many minutes will it take her to run three miles? (1 mile = 4 laps)

2. Jamar needs to be at school at 7:45 A.M. It takes him 25 minutes to walk to school, 25 minutes to eat, and 35 minutes to shower and get dressed. What time should he get up to be at school 5 minutes early?

Summary

Write 2–3 sentences to summarize the lesson.

Rate Yourself!

How well do you understand the different problem-solving strategies? Circle the image that applies.

Clear   Somewhat Clear   Not So Clear

For more help, go online to access a Personal Tutor.

Lesson 1-6

Ordered Pairs and Relations

Getting Started

Scan Lesson 1-6 in your textbook. List two real-world scenarios in which you would use a coordinate system.

- _____
  _____

- _____
  _____

Vocabulary
Write the definition of *coordinate system* in your own words.

_____

_____

_____

_____

Vocabulary Start-Up

A set of ordered pairs such as {(1, 2), (2, 4), (3, 0), (4, 5)} is a **relation**. The **domain** of the relation is the set of *x*-coordinates. The **range** of the relation is the set of *y*-coordinates.

Label the sets below with the terms *domain, ordered pair, range,* and *relation*.

$$\{(1, 2), (2, 4), (3, 0), (4, 5)\}$$

$$\{1, 2, 3, 4\} \qquad \{0, 2, 4, 5\}$$

Real-World Link

Bungee Jumping The table describes the approximate heights and times of the jumps at four bungee jumping sites.

Height (ft)	Time of Fall (s)
170	3
452	5
630	6
764	7

1. Write the data as a relation with ordered pairs (feet, seconds).

   _____

2. Write the domain and range of the relation.

   domain: _____

   range: _____

Notes

Ordered Pairs

Graph each ordered pair on the coordinate plane.

1. $A(4, 7)$

2. $B(6, 0)$

3. $C(2, 5)$

4. $D(0, 3)$

5. $E(7, 4)$

6. $F(5, 2)$

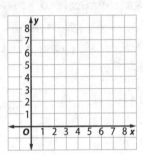

7. How would you explain to a friend the steps for graphing a point on the coordinate plane? _____

_____

Relations

Express each relation as a table. Then write the domain and range.

8. $\{(2, 2), (4, 1), (3, 0), (6, 6)\}$

x				
y				

   domain: _____

   range: _____

9. $\{(3, 4), (2, 1), (1, 2), (4, 5)\}$

x			
y			

   domain: _____

   range: _____

Summary

Write 2–3 sentences to summarize the lesson.

_____

_____

_____

_____

_____

_____

Rate Yourself!

Are you ready to move on? Shade the section that applies.

- I have a few questions.
- I'm ready to move on.
- I have a lot of questions.

For more help, go online to access a Personal Tutor.

Tutor

Words, Equations, Tables, and Graphs

Getting Started

Scan Lesson 1-7 in your textbook. Predict two things you will learn about multiple representations of equations.

- _____
  _____
  _____

- _____
  _____
  _____

Vocabulary

Write the definition of *equation* in your own words.

_____

_____

_____

 **Real-World Link**

Fireworks Physics can be used to calculate the path of fireworks. In general, for every 1-inch increase in shell diameter, a firework's height increases by about 100 feet.

1. How does the height of a firework relate to the diameter of its shell?

_____

_____

2. Complete the table to show the approximate height of a firework with different sized shells. Then graph the data as ordered pairs (diameter, height).

Fireworks	
Diameter (in.)	Height (ft)
1	100
2	
3	
4	

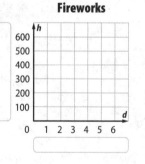

Fireworks

3. How could you find the approximate height of a firework

with any sized shell? _____

_____

4. Based on the rule you wrote in Exercise 3, what is the approximate height of a firework with a 6-inch shell?

_____

an 8-inch shell? _____

Notes

Represent Relations

A store rents scooters for $12.25 per hour as shown in the table. In the graphic organizer, write this relationship using words and an equation.

Time (h)	1	2	3	4
Total Cost ($)	12.25	24.50	36.75	49.00

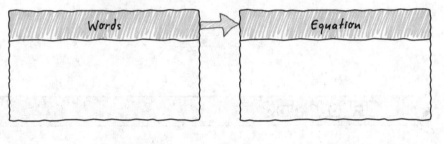

Words   Equation

Multiple Representations

The height *y* of a hot air balloon increases by 20 feet every minute after it lifts off the ground *x*. Use a table, a graph, and an equation to represent this situation.

Table

Time (min)					
Height (ft)					

Equation _____

Graph

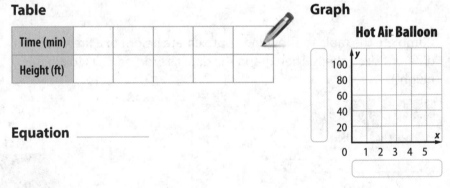

Hot Air Balloon

Summary

Write 2–3 sentences to summarize the lesson.

_____

_____

_____

_____

_____

_____

_____

Rate Yourself!

How confident are you about using multiple representations of relations? Check the box that applies.

☹   😐   😊

☐   ☐   ☐   ☐   ☐

For more help, go online to access a Personal Tutor.

Tutor

Chapter Review

Vocabulary Check

Complete the crossword puzzle using the vocabulary list at the beginning of the chapter.

Across

1. statements that are true for any numbers

5. the set of all *x*-values in a relation

6. a type of expression that contains at least one variable

8. the type of reasoning that uses facts, properties, or rules to reach a valid conclusion

Down

2. the point at which the number lines on a coordinate plane intersect

3. a mathematical sentence stating that two quantities are equal

4. a letter or symbol used to represent an unknown

7. the rules to follow when evaluating an expression are called the ___?___ of operations

Use Your FOLDABLES

Use your Foldable to help review the chapter.

Tape here

Expressions and Properties

To evaluate a numerical expression:

To evaluate an algebraic expression:

To evaluate an expression using properties:

Got it?

The problems below may or may not contain an error. If the problem is correct, write a "✓" by the answer. If the problem is not correct, write an "X" over the answer and correct the problem.

Evaluate each expression if $a = 6$, $b = 3$, and $c = 11$.

1. $2a + 4c$; 56

2. $\frac{10b}{a}$; 24

3. $2c + 3b$; 56

Problem Solving

1. A 500-gallon bathtub is being filled with water. Eighty gallons of water are in the bathtub after 4 minutes. How long will it take to fill the bathtub?

   (Lessons 1 and 5) _____

2. The Spanish Club is hosting a luncheon for all the students. They are charging $1.00 for a taco, $3.00 for a burrito, and $0.75 for a drink. Michael is buying lunch for his friends. If he buys 3 tacos, 2 burritos and 3 drinks, write and solve an

   expression to find the total cost. (Lesson 2) _____

3. Admission to a high school football game is $8. At the game, raffle tickets are sold for $0.50 each. Write an expression showing the cost of admission and a purchase of t tickets. Then find the total cost if 20 tickets are purchased.

   (Lesson 3) _____

4. After school, Anisha usually has a snack and works on her homework. Are the

   actions commutative? Explain. (Lesson 4) _____

5. Pandas eat about 240 pounds of bamboo every three days. (Lessons 6 and 7)

   a. Write an equation that can be used to find the pounds of bamboo p a panda

   will eat in any number of days d. _____

   b. Make a table of values to find the pounds of bamboo a panda will eat in 5, 7, 10, and 13 days. Then graph the ordered pairs (d, p).

Days (d)	80d	Pounds (p)

Bamboo Consumption

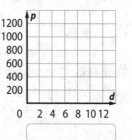

Reflect

 Answering the Essential Question

Use what you learned about algebric expressions to complete the graphic organizer. Then answer the chapter's Essential Question below.

When do you use a variable?

 Essential Question

How can you use numbers and symbols to represent mathematical ideas?

How do you know which operation symbol to use?

Answer the Essential Question. HOW can you use numbers and symbols to represent mathematical ideas?

Chapter 2

Operations with Integers

Chapter Preview

Vocabulary

absolute value	inductive reasoning	negative number	quadrant
additive inverse	inequality	opposites	zero pair
coordinate	integer	positive number	

Vocabulary Activity

Use the Glossary to find the definitions of the terms below. Then draw a line to match each term with the correct definition.

1. negative number

2. integers

3. opposites

4. coordinate

5. inequality

6. absolute value

7. zero pair

8. additive inverses

9. inductive reasoning

10. quadrant

a. A mathematical sentence that contains $<, >, \neq, \leq,$ or $\geq$.

b. The distance a number is from zero on a number line.

c. One of four regions into which the x-axis and y-axis separate the coordinate plane.

d. A number less than zero.

e. An integer and its opposite.

f. Making a conjecture based on a pattern of examples or past events.

g. Two numbers with the same absolute value but different signs.

h. A number that corresponds to a point on a number line.

i. The set of whole numbers and their opposites.

j. A positive tile paired with a negative tile.

CCSS **Quick Review**   **Common Core Review** 6.EE.2, 6.NS.6

Example 1

Evaluate $2b + 3c$ if $b = 5$ and $c = 6$.

$$2b + 3c = 2(5) + 3(6) \quad \text{Replace } b \text{ with 5 and } c \text{ with 6.}$$
$$= 10 + 18 \quad \text{Multiply.}$$
$$= 28 \quad \text{Add.}$$

Example 2

Use the coordinate plane to write the ordered pair that names point A.

Step 1 Start at the origin.

Step 2 Move right on the x-axis to find the x-coordinate of point A, which is 4.

Step 3 Move up the y-axis to find the y-coordinate, which is 1.

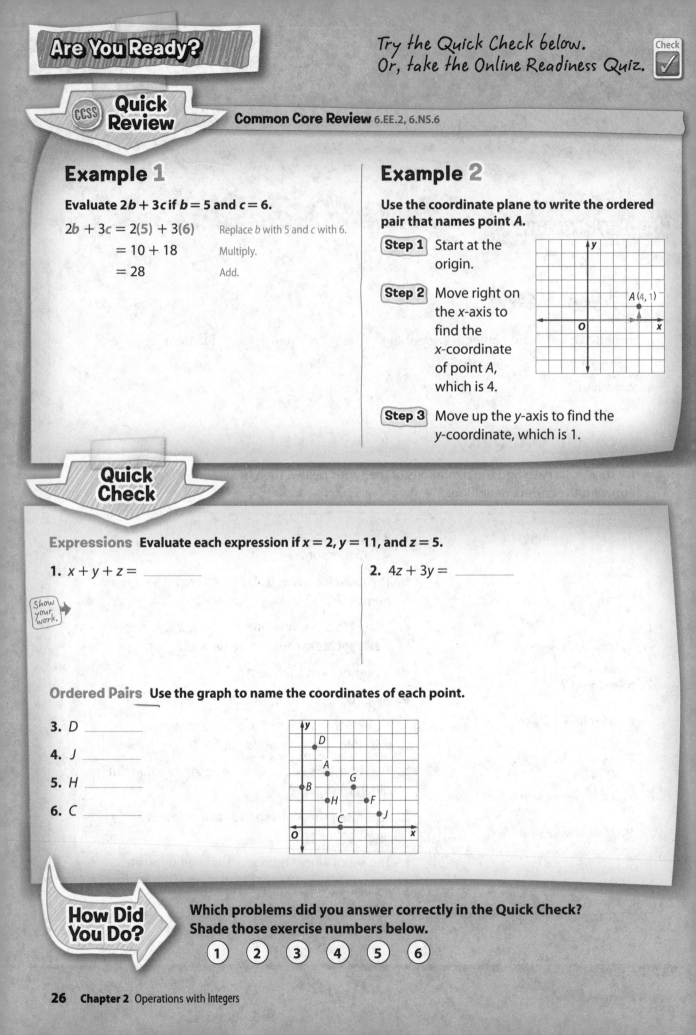

Quick Check

Expressions **Evaluate each expression if $x = 2$, $y = 11$, and $z = 5$.**

1. $x + y + z =$ _____

2. $4z + 3y =$ _____

Show your work.

Ordered Pairs **Use the graph to name the coordinates of each point.**

3. D _____

4. J _____

5. H _____

6. C _____

How Did You Do?

Which problems did you answer correctly in the Quick Check?
Shade those exercise numbers below.

① ② ③ ④ ⑤ ⑥

cut on all dashed lines   fold on all solid lines   tape to page 44   FOLDABLES

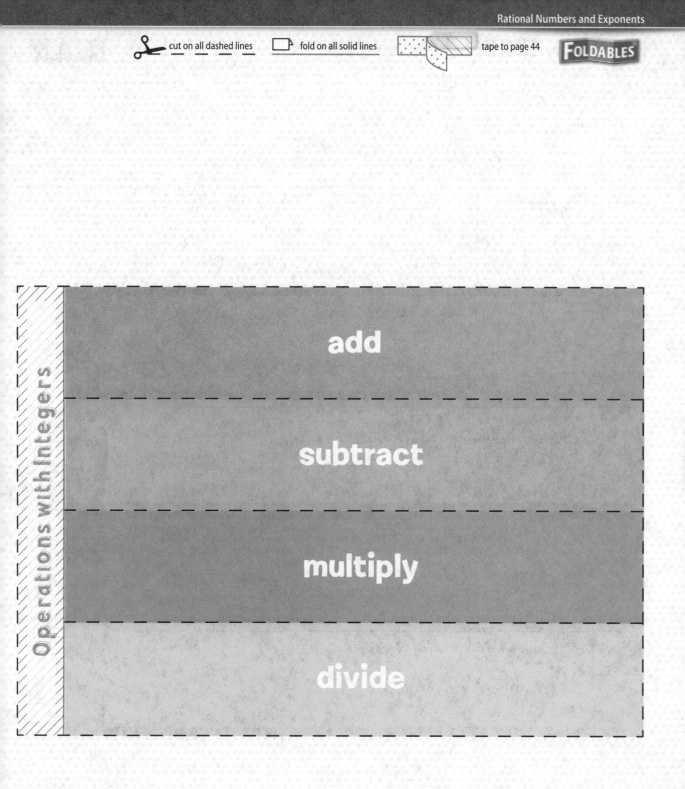

Operations with Integers

add

subtract

multiply

divide

FOLDABLES®
Study Organizer

1 Cut out the Foldable above.

2 Place your Foldable on page 44.

3 Use the Foldable throughout this chapter to help you learn about operations with integers.

How do I add integers with the same sign?

+

How do I subtract integers with the same sign?

−

How do I multiply integers with the same sign?

×

How do I divide integers with the same sign?

÷

page 44

Integers and Absolute Value

Getting Started

Scan Lesson 2-1 in your textbook. List two headings you would use to make an outline of the lesson.

- _____
  _____
  _____
- _____
  _____

Quick Review

Graph {1, 2, 4} on the number line below.

0 1 2 3 4

Vocabulary Start-Up

Numbers like 8 and −7 are called integers. An **integer** is any number from the set {…, −3, −2, −1, 0, 1, 2, 3, …}, where … means *continues indefinitely*.

Complete the graphic organizer.

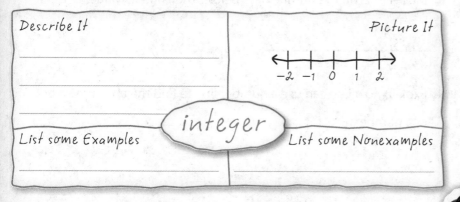

Describe It

Picture It

−2 −1 0 1 2

integer

List some Examples

List some Nonexamples

Real-World Link

Geocaching *Geocaching* is an outdoor treasure hunting game. Some treasures, or *geocaches*, are located thousands of feet above sea level. Their locations can be represented by positive integers. Other geocaches are hidden in lakes. Their locations below the surface can be represented by negative integers.

Circle the integer that best represents each location.

1. 2972 feet above sea level on Mt. Wilson Trail, California

   2972   −2972

2. 33 feet deep in Lake Denton, Florida

   33   −33

Notes

Compare and Order Integers

Fill in each ◯ with < or > to make a true mathematical sentence.

1. 0 ◯ −3

2. −10 ◯ 0

3. −1 ◯ −4

4. −19 ◯ −17

Because of a drought, the water level of a river is 3 feet below its normal level. Use this information for Exercises 5–7.

5. The integer _____ represents this situation.

6. The opposite of this integer is _____.

7. What does the opposite of the original integer mean?

_____

Absolute Value

8. Model $|-2|$ on the number line below. Then find its value.

 $|-2| = \boxed{}$

9. Explain how you can use a number line to find the absolute value

of a number. _____

_____

_____

Summary

Write 2–3 sentences to summarize the lesson.

_____

_____

_____

_____

_____

_____

Rate Yourself!

How confident are you about integers and absolute value? Shade the ring on the target.

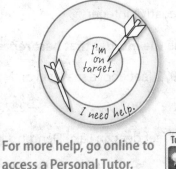

I'm on target.

I need help.

For more help, go online to access a Personal Tutor.

Lesson 2-2
Adding Integers

Getting Started

Scan Lesson 2-2 in your textbook. Predict two things you will learn about adding integers.

- _____
  _____
  _____

- _____
  _____
  _____

Vocabulary

Write the definition of *opposites* in your own words.

_____

_____

_____

Real-World Link

Financial Literacy The money you earn can be represented by a positive integer, while the money you spend or borrow can be represented by a negative integer. Mei borrowed $15 from her parents on Friday night to go to a movie with her friends. On Saturday, she earned $25 babysitting for her neighbors.

1. Write an integer to represent each situation.

   Mei owes her parents $15. ☐

   Mei earned $25 babysitting. ☐

2. If Mei started with $0, what addition expression could be used to find the amount of money she has after repaying her parents

   and babysitting? _____

3. Fill in the number line to show $-15 + 25$. Then write an addition sentence to represent the number line.

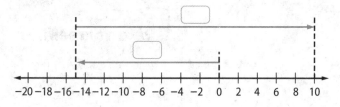

   _____

4. Suppose Mei spent another $7 after repaying her parents. What addition sentence would show the amount of money she has left?

   _____

Notes

Add Integers

Model the addition sentence 3 + (−4) on the number line below and find the sum.

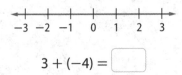

$$3 + (-4) = \boxed{}$$

Circle the word that makes each statement true.

1. To add a positive number, you move (right, left) on a number line.

2. To add a negative number, you move (right, left) on a number line.

Add More than Two Integers

Fill in each blank with the property used to simplify the expression.

$5 + (-7) + 2$

$= 5 + 2 + (-7)$ _____

$= (5 + 2) + (-7)$ _____

$= 7 + (-7)$ Simplify.

$= 0$ _____

Summary

Write 2–3 sentences to summarize the lesson.

_____

_____

_____

_____

_____

_____

_____

Rate Yourself!

Are you ready to move on? Shade the section that applies.

YES   ?   NO

For more help, go online to access a Personal Tutor.

Tutor

FOLDABLES Time to update your Foldable!

Lesson 2-3
Subtracting Integers

Getting Started

Scan Lesson 2-3 in your textbook. List two real-world scenarios in which you would subtract integers to solve.

- _____

  _____

- _____

  _____

Real-World Link

Ants Ant colonies can be found both above and below ground. Most colonies build their nests below ground, where they can dig to 20 feet below the surface. Other ant colonies build above ground nests that can sometimes reach 6 feet tall.

1. What integer would you use to represent *20 feet below the surface*? Explain your reasoning. _____

   _____

   _____

2. What integer would you use to represent *6 feet tall*? Explain.

   _____

   _____

3. Graph the integers that you wrote in Exercises 1 and 2 on the number line at the right.

4. How many units apart are the two points? _____

5. What does this distance represent?

   _____

   _____

   _____

6. Write a subtraction sentence to represent the distance between the points that you graphed.

   _____

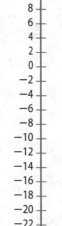

Notes

Subtract Integers

Fill in the boxes to show how addition is used to subtract integers.

1. $3 - 4$ ·········▶ $3 + \boxed{} = \boxed{}$

2. $-1 - 6$ ·········▶ $-1 + \boxed{} = \boxed{}$

3. $-2 - (-3)$ ·········▶ $-2 + \boxed{} = \boxed{}$

Find Distance on a Number Line

Determine whether each expression or phrase is represented on the number line below. Circle Y for *yes* or N for *no*.

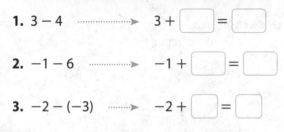

4. $3 - 4$ Y   N

5. $-4 - 3$ Y   N

6. $3 - 7$ Y   N

7. the distance between -4 and 3   Y   N

8. the distance between 3 and 7   Y   N

Summary

Write 2–3 sentences to summarize the lesson.

_____

_____

_____

_____

_____

_____

Rate Yourself!

How confident are you about subtracting integers? Check the box that applies.

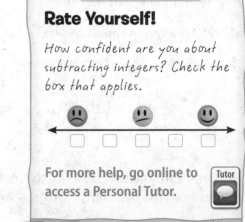

For more help, go online to access a Personal Tutor.

FOLDABLES *Time to update your Foldable!*

Mid-Chapter Check

Vocabulary Check

1. **CCSS Identify Structure** Define *additive inverses*. Give two examples of additive inverses. (Lesson 2) _____

_____

2. Fill in the blank in the sentence below with the correct term. (Lesson 1)

A(n) _____ is any number from the set $\{\ldots, -3, -2, -1, 0, 1, 2, 3, \ldots\}$.

Skills Check and Problem Solving

Write an integer for each situation. Then graph the integer on a number line. (Lesson 1)

3. 300 feet below sea level _____

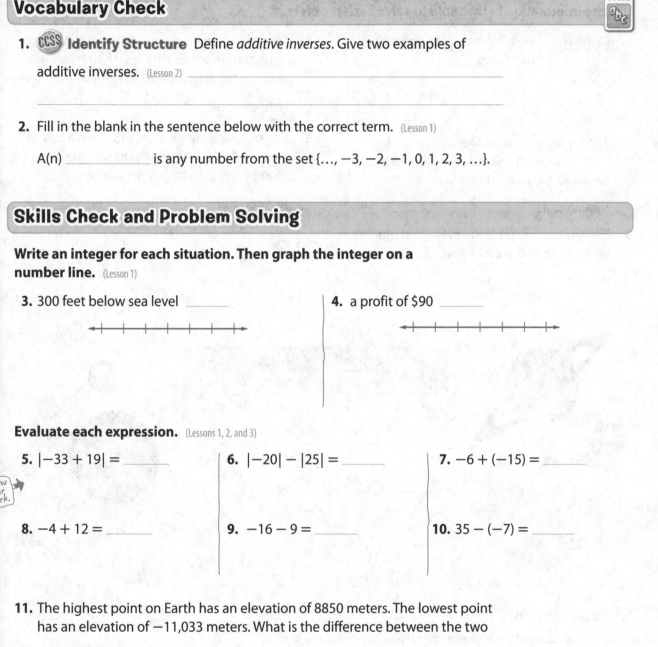

4. a profit of $90 _____

Evaluate each expression. (Lessons 1, 2, and 3)

5. $|-33 + 19| =$ _____

6. $|-20| - |25| =$ _____

7. $-6 + (-15) =$ _____

8. $-4 + 12 =$ _____

9. $-16 - 9 =$ _____

10. $35 - (-7) =$ _____

11. The highest point on Earth has an elevation of 8850 meters. The lowest point has an elevation of $-11,033$ meters. What is the difference between the two elevations? (Lesson 3) _____

12. **Standardized Test Practice** Refer to the number line below.

Which statement is true? (Lesson 1)

Ⓐ $|B| < |C|$　　　Ⓑ $B > C$　　　Ⓒ $C > A$　　　Ⓓ $|D| > |A|$

Predicting Space Storms!

Use the information in the table to solve each problem.

1. Order the planets from least to greatest average temperatures.

   _____

   _____

2. The temperatures on Mercury range from −279°F to 800°F. What is the difference between the highest and lowest

   temperatures? _____

3. How much greater is the average temperature on Earth than the average temperature on

   Jupiter? _____

4. One of Neptune's moons, Triton, has a surface temperature that is 61°F less than Neptune's average temperature. What is Triton's surface

   temperature? _____

5. The temperature on Mars can reach a low of about −225°F. Is this temperature greater than or less than Neptune's average

   temperature? _____

Average Temperature of Planets (°F)			
Planet	Temperature	Planet	Temperature
Earth	59	Neptune	−330
Jupiter	−166	Saturn	−220
Mars	−85	Uranus	−320
Mercury	333	Venus	867

Career Project

It's time to update your career portfolio! Investigate the education and training requirements for a career as a space weather forecaster.

_____

_____

_____

_____

_____

List other careers that someone with an interest in astronomy could pursue.

- _____
- _____
- _____
- _____
- _____

Multiplying Integers

Getting Started

Scan Lesson 2-4 in your textbook. List two headings you would use to make an outline of the lesson.

- _____
  _____

- _____
  _____

Real-World Link

Ballooning A hot air balloon pilot is flying a balloon 560 meters above the ground. When landing, the balloon descends at a rate of about 8 meters per second.

1. Would you use a positive or negative integer to represent the balloon's rate of change? Explain your reasoning. _____

   _____

2. Complete the table below to show how far the balloon will descend after 5 seconds.

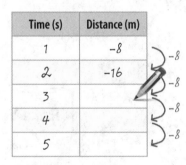

Time (s)	Distance (m)
1	−8
2	−16
3	
4	
5	

3. Write an addition expression to represent the total change in the balloon's height after 5 seconds. What is the sum?

   _____

4. Multiplication is defined as repeated addition. Rewrite the expression you wrote in Exercise 3 as a multiplication expression. _____

5. Predict the value of −8(5). _____

6. How long will it take for the balloon to land? _____

Notes

Multiply Integers

Circle the sign of the product.

1. positive · positive   +   −

2. positive · negative   +   −

3. negative · negative   +   −

4. negative · positive   +   −

5. negative · negative · negative   +   −

6. positive · negative · positive   +   −

Algebraic Expressions

Use the reason given for each step to simplify each expression.

7. $(-6)(4)(-3) =$ _____   Associative Property

            $=$ _____   Simplify.

            $=$ _____   Simplify.

8. $-3(10m)(2p) =$ _____   Commutative Property

            $=$ _____   Associative Property

            $=$ _____   Simplify.

            $=$ _____   Simplify.

Summary

Write 2–3 sentences to summarize the lesson.

_____

_____

_____

_____

_____

_____

Rate Yourself!

How well do you understand multiplying integers? Circle the image that applies.

Clear   Somewhat Clear   Not So Clear

For more help, go online to access a Personal Tutor.

Tutor

FOLDABLES *Time to update your Foldable!*

Lesson 2-5

Dividing Integers

Getting Started

Scan Lesson 2-5 in your textbook. Predict two things you will learn about dividing integers.

- _____
  _____

- _____
  _____

Vocabulary

Write the definition of *mean* in your own words.

_____

_____

_____

_____

Real-World Link

Hair The number of hairs you have on your head depends on several different factors. Age, gender, and hair color are just a few of the factors. The average person loses about 560 hairs per week.

1. What integer represents the average person's change in the number of hairs per week? [____]

2. How could you find the number of hairs the average person loses each day?

   _____

   _____

3. What expression could be used to find the number of hairs lost every day? _____

4. Will the quotient be positive or negative? Explain your reasoning.

   _____

   _____

   _____

5. What integer represents the average person's daily change in the number of hairs? _____

6. The table shows the approximate number of hairs a person has based on their hair color. Suppose a person does not regrow any hair. About how many weeks would it take for them to lose all of the hair on their head? _____

   _____

Hair Color	Average Number of Hairs
blonde	120,000
brown	110,000
black	105,000
red	80,000

Notes

Divide Integers

Write *positive* or *negative* for each quotient. Then find the quotient.

$-21 \div 7$	$-18 \div (-3)$
$24 \div 6$	$15 \div (-5)$

Mean (Average)

In five rounds of golf, Evan scored +4, +1, −2, +3, and −1. Fill in the boxes to find Evan's mean score.

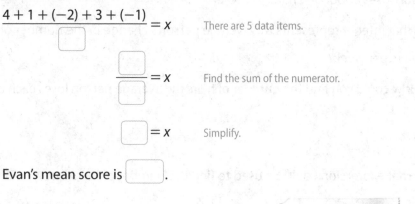

$\dfrac{4 + 1 + (-2) + 3 + (-1)}{\boxed{}} = x$ There are 5 data items.

$\dfrac{\boxed{}}{\boxed{}} = x$ Find the sum of the numerator.

$\boxed{} = x$ Simplify.

Evan's mean score is $\boxed{}$.

Summary

Write 2–3 sentences to summarize the lesson.

Rate Yourself!

☐ I understand how to divide integers.

▶▶ **Great! You're ready to move on!**

☐ I still have questions about dividing integers.

▌▌ **No Problem! Go online to access a Personal Tutor.** Tutor 💬

FOLDABLES *Time to update your Foldable!*

Graphing in the Four Quadrants

Getting Started

Scan Lesson 2-6 in your textbook. List two headings you would use to make an outline of the lesson.

- _____
  _____

- _____
  _____

Vocabulary

Circle the vocabulary word defined below.

One of four regions into which the *x*-axis and *y*-axis separate the coordinate plane.

origin   quadrant

Vocabulary Start-Up

The coordinate system can be extended to include points below and to the left of the origin.

Label the coordinate plane with the terms *ordered pair, origin, x-axis, x-coordinate, y-axis,* **and** *y-coordinate.*

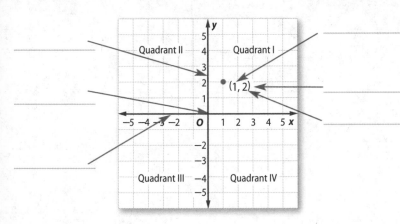

Real-World Link

Video Games Programmers of 3-D video games use coordinate systems or *spaces* to create a game engine. Describe how objects in a video game could be placed using a coordinate system.

_____

_____

_____

Notes

Graph Points

Draw a line to connect each "ordered pair" to the correct quadrant.

1. $(-, +)$

2. $(+, -)$

3. $(+, +)$

4. $(-, -)$

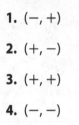

Graph Algebraic Relationships

Complete the table and graph the points for the following relationship.

The sum of a negative number and a positive number is 0.

Rule: $x + y = 0$	
x	y
-4	
-3	
-2	
-1	

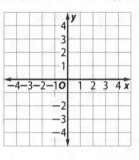

Summary

Write 2–3 sentences to summarize the lesson.

_____

_____

_____

_____

_____

_____

Chapter Review

Vocabulary Check

Unscramble each of the clue words. After unscrambling all of the terms, use the numbered letters to find another vocabulary term from the chapter.

SANQTDUAR

[ ][ ][ ][ ][ ][ ][ ][ ][ ]
　　　　　3

SIVPITEO MBNURE

[ ][ ][ ][ ][ ][ ][ ][ ]　　[ ][ ][ ][ ][ ][ ]
　　　　　　　2

LOATSBEU VUELA

[ ][ ][ ][ ][ ][ ][ ][ ]　　[ ][ ][ ][ ][ ]
　　　8　　　　　　　　　　6

TAIEIDVD ENEIVRS

[ ][ ][ ][ ][ ][ ][ ][ ]　　[ ][ ][ ][ ][ ][ ][ ]
　　　　1　　　　　　　　　4

RNADOEOCTI

[ ][ ][ ][ ][ ][ ][ ][ ][ ]
　　　7

GVEAITNE UNMREB

[ ][ ][ ][ ][ ][ ][ ][ ]　　[ ][ ][ ][ ][ ][ ]
　　　5

[ ][ ][ ][ ][ ][ ][ ][ ]
1　2　3　4　5　6　7　8

Complete each sentence using one of the unscrambled words above.

1. The _____ of a number is the distance the number is from zero on the number line.

2. A(n) _____ number is a number that is less than zero.

3. An integer and its opposite can also be called _____.

4. The _____ is the number that corresponds to a point on a number line.

5. A(n) _____ number is a number that is greater than zero.

6. The *x*-axis and *y*-axis separate the coordinate plane into four _____.

Use Your FOLDABLES

Use your Foldable to help review the chapter.

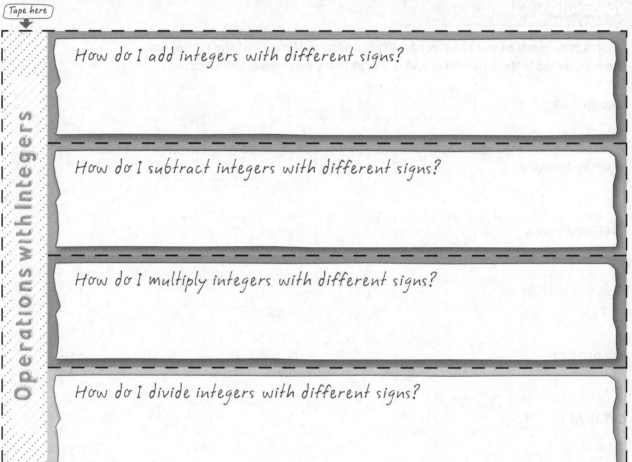

Tape here

Operations with Integers

How do I add integers with different signs?

How do I subtract integers with different signs?

How do I multiply integers with different signs?

How do I divide integers with different signs?

Got it?

Draw a line to match each expression with its value.

1. $-8 + 12$ **a.** -4

2. $-4 - 6$ **b.** 10

3. $-16 \div 4$ **c.** 4

4. $(-2)(-5)$ **d.** 12

5. $8 - (-4)$ **e.** -10

Problem Solving

1. The table shows the maximum surface temperatures for different planets. Order the temperatures from least to greatest. (Lesson 1)

Show your work.

Planet	Maximum Surface Temperature (°C)
Earth	58
Mars	−5
Mercury	427
Neptune	−214
Jupiter	−148
Uranus	−216

2. Kyle owes his sister $24. He gives his sister $15 that he earned mowing lawns. Write and simplify an addition expression to find how much Kyle still owes his sister. (Lesson 2) _____

3. A commercial jet has a usual cruising altitude of 34,000 feet. Because of a storm, it climbs 2500 feet and then descends 4700 feet to its new cruising altitude. What is the jet's new cruising altitude? (Lesson 3) _____

4. The wearing away of shoreline, or *erosion rate*, at Emerald Isle in North Carolina is about 2 feet per year. If erosion continues at the current rate, what integer represents the change in coastline at Emerald Isle after 13 years? (Lesson 4)

5. After 6 seconds, the Dynamic Drop is 18 feet below the top of the ride. What integer represents the average change in the ride's location in feet per second?

(Lesson 5) _____

6. The difference between two golf scores is 4. If x represents one score and y represents the other score, complete the table. Then graph the ordered pairs and describe the graph. (Lesson 6)

$x - y = 4$		
x	y	(x, y)
−2		
	−5	
0		
	−3	
2		

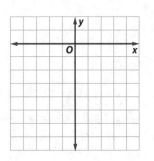

Reflect

 Answering the Essential Question

Use what you learned about integers to complete the graphic organizer. Explain how to determine the sign of the result when performing each operaton.

Add and Subtract

Essential Question

WHAT happens when you add, subtract, multiply, and divide integers?

Multiply and Divide

Answer the Essential Question. WHAT happens when you add, subtract, multiply, and divide integers?

Chapter 3

Operations with Rational Numbers

Chapter Preview

 Vocabulary

bar notation

like fractions

multiplicative inverse

rational number

reciprocal

repeating decimal

terminating decimal

unlike fractions

Vocabulary Activity

Select three vocabulary terms from the list above. Write a definition for the terms based upon what you already know. As you go through the chapter, come back to this page and update your definitions if necessary.

Term	What I Know	What I Learned

Are You Ready?

Try the Quick Check below.
Or, take the Online Readiness Quiz.

Check ✓

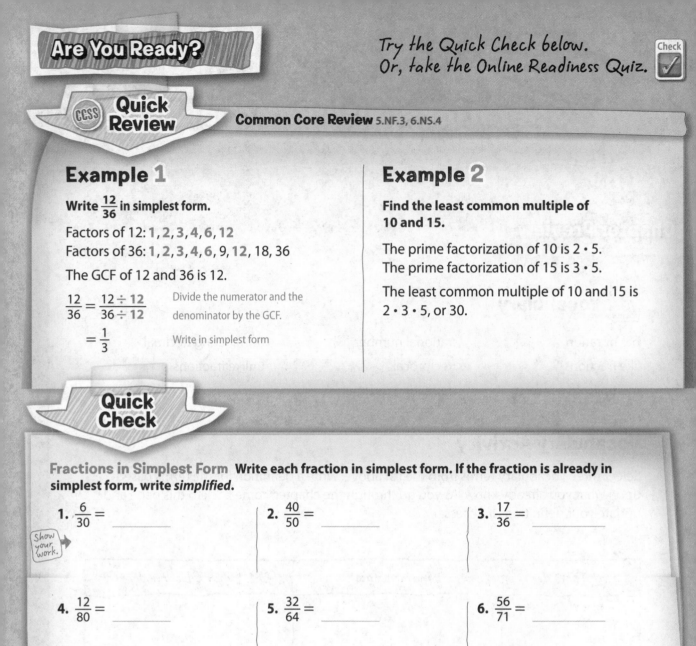

CCSS Quick Review

Common Core Review 5.NF.3, 6.NS.4

Example 1

Write $\frac{12}{36}$ in simplest form.

Factors of 12: **1, 2, 3, 4, 6, 12**
Factors of 36: **1, 2, 3, 4, 6,** 9, **12,** 18, 36

The GCF of 12 and 36 is 12.

$\frac{12}{36} = \frac{12 \div 12}{36 \div 12}$ Divide the numerator and the denominator by the GCF.

$= \frac{1}{3}$ Write in simplest form

Example 2

Find the least common multiple of 10 and 15.

The prime factorization of 10 is 2 • 5.
The prime factorization of 15 is 3 • 5.

The least common multiple of 10 and 15 is 2 • 3 • 5, or 30.

Quick Check

Fractions in Simplest Form Write each fraction in simplest form. If the fraction is already in simplest form, write *simplified*.

Show your work.

1. $\frac{6}{30} =$ _____

2. $\frac{40}{50} =$ _____

3. $\frac{17}{36} =$ _____

4. $\frac{12}{80} =$ _____

5. $\frac{32}{64} =$ _____

6. $\frac{56}{71} =$ _____

Least Common Multiple Find the least common multiple for each pair of numbers.

7. 3, 5 _____

8. 6, 8 _____

9. 12, 36 _____

10. 10, 50 _____

11. 14, 21 _____

12. 30, 40 _____

How Did You Do?

Which problems did you answer correctly in the Quick Check?
Shade those exercise numbers below.

① ② ③ ④ ⑤ ⑥ ⑦ ⑧ ⑨ ⑩ ⑪ ⑫

✂ cut on all dashed lines   ▢ fold on all solid lines   tape to page 66   **FOLDABLES**

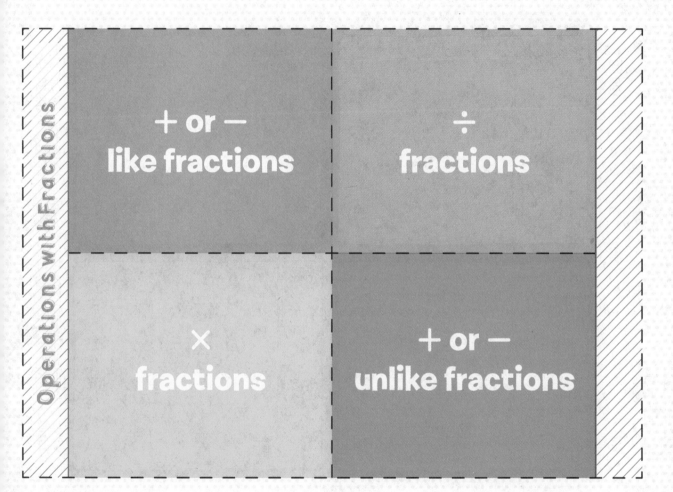

Operations with Fractions

+ or −
like fractions

÷
fractions

×
fractions

+ or −
unlike fractions

FOLDABLES Study Organizer

1 Cut out the Foldable above.

2 Place your Foldable on page 66.

3 Use the Foldable throughout this chapter to help you learn about operations with fractions.

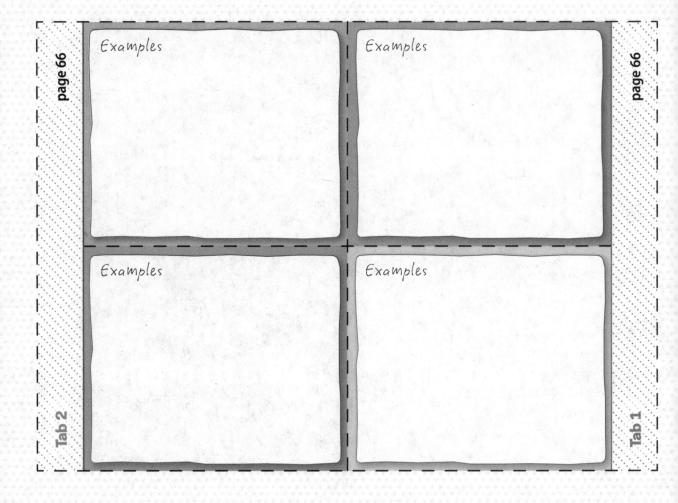

page 66

Examples

Examples

page 66

Tab 2

Examples

Examples

Tab 1

Lesson 3-1

Fractions and Decimals

Getting Started

Scan Lesson 3-1 in your textbook. Predict two things you will learn about fractions and decimals.

- _____
- _____
- _____
- _____

Quick Review

Write the following decimals as fractions in simplest form.

0.4, 0.25

Vocabulary Start-Up

A **repeating decimal** has a pattern in its digits that repeats without end and can be written using bar notation. **Bar notation** is a bar or line placed over the digit(s) that repeat. For example, the repeating decimal 0.12525… can be written as $0.1\overline{25}$.

Write an X through the number that has a different value than the two remaining numbers.

1. $0.2\overline{7}$, 0.27, 0.2777…

2. 0.39, $0.3\overline{9}$, 0.3999…

3. 0.161616…, $0.\overline{16}$, 0.16

4. $3.\overline{4}$, 3.4, 3.444…

Real-World Link

Robotics In an annual robot competition, middle school students apply math and science to design, program, and test their own robots. The goal is to make their 'bots outperform the competition.

Circle the correct decimal length of each robotics part.

5. patch cable: $\frac{3}{10}$-meter   0.03 m or 0.3 m

6. battery cable: $\frac{1}{2}$-meter   0.5 m or 0.2 m

7. power board: $\frac{1}{5}$-foot   0.5 ft or 0.2 ft

8. red power wire: $\frac{3}{4}$-yard   3.4 yd or 0.75 yd

Notes

Write Fractions as Decimals

Complete the model to write $\frac{13}{20}$ as a decimal.

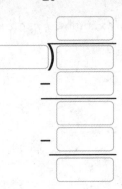

In your own words, how do you write a fraction as a decimal?

_____

_____

Compare Fractions and Decimals

Fill in the steps necessary to replace the ◯ with <, >, or =.

$$\frac{11}{20} \bigcirc 0.65$$

Step 1	Write $\frac{11}{20}$ as a _____.	$\frac{11}{20} = $ ___
Step 2	Compare the tenths place.	___ ◯ 0.6
Step 3	Write a true statement.	$\frac{11}{20}$ ◯ 0.65

Summary

Write 2–3 sentences to summarize the lesson.

_____

_____

_____

_____

_____

_____

Rate Yourself!

Are you ready to move on? Shade the section that applies.

I have a few questions.

I'm ready to move on.

I have a lot of questions.

For more help, go online to access a Personal Tutor.

Tutor

Rational Numbers

Getting Started

Scan Lesson 3-2 in your textbook. List two headings you would use to make an outline of the lesson.

- _____

  _____

- _____

  _____

Vocabulary Start-Up

Any number that can be written as a fraction is part of the set of **rational numbers**.

Complete the graphic organizer by writing each rational number as a fraction.

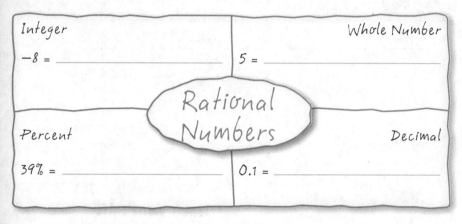

Integer	Whole Number
−8 = _____	5 = _____

Rational Numbers

Percent	Decimal
39% = _____	0.1 = _____

 ## Real-World Link

Monkeys New World monkeys are small to mid-sized, they possess high intelligence, and they are skilled in using their hands. *Spider monkeys* are one type of New World monkey. Identify the sets to which each number that decribes a spider monkey belongs. Write *whole, integer*, and/or *rational*.

1. A spider monkey's diet is $\frac{9}{10}$ fruit. _____

2. The average lifespan of a spider monkey is 30 years. _____

3. Black-headed spider monkeys weigh about 10.8 kilograms. _____

Notes

Rational Numbers

Match each repeating decimal with its equivalent fraction.

1. $0.\overline{3}$ $\dfrac{14}{33}$

2. $0.\overline{64}$ $\dfrac{1}{3}$

3. $0.\overline{42}$ $\dfrac{1}{33}$

4. $0.\overline{03}$ $\dfrac{64}{99}$

Identify and Classify Rational Numbers

Identify the set(s) to which each number belongs by writing an "X" in the appropriate column(s). The first one is done for you.

Number	Natural	Whole	Integer	Rational
-14			X	X
7				
0.25				
0				
$\dfrac{6}{7}$				

Summary

Write 2–3 sentences to summarize the lesson.

_____

_____

_____

_____

_____

_____

Rate Yourself!

How confident are you about with rational numbers? Shade the ring on the target.

I'm on target.

I need help.

For more help, go online to access a Personal Tutor.

Multiplying Rational Numbers

Getting Started

Scan Lesson 3-3 in your textbook. List two real-world scenarios in which you would solve a problem by multiplying fractions or mixed numbers.

- _____
  _____

- _____
  _____

> **Quick Review**
>
> What is the greatest common factor (GCF) of 4 and 10?
>
> _____

🌎 Real-World Link

Ice Cream A survey was taken to find the most popular ice cream flavors in the United States. The top ten flavors are shown below.

Rank	Flavor	Rank	Flavor
1	vanilla	6	chocolate chip
2	chocolate	7	French vanilla
3	butter pecan	8	cookies and cream
4	strawberry	9	vanilla fudge ripple
5	Neapolitan	10	praline pecan

1. What fraction of the flavors contains chocolate? (*Hint:* Neapolitan contains chocolate, vanilla, and strawberry.) _____

2. What fraction of the flavors in Exercise 1 also contain strawberry?

   _____

3. Shade the model below to show the fraction in Exercise 1. Then shade the model to show the fraction in Exercise 2.

4. What fraction of the model is shaded twice? _____

   What does this mean? _____

   _____

Notes

Multiply Fractions

Find the product of the center fraction and the number in each section. Write each product in simplest form.

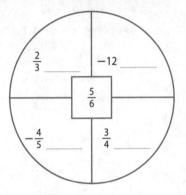

Evaluate Expressions with Fractions

Fill in the blanks to find each product in simplest form if $w = \frac{2}{3}$, $y = -\frac{7}{11}$, and $z = \frac{3}{5}$.

1. wy

$\boxed{} \times \boxed{} = \boxed{}$

2. $-2y$

$-2 \times \boxed{} = \boxed{}$

3. $\frac{5}{9}z$

$\frac{5}{9} \times \boxed{} = \boxed{}$

4. wyz

$\boxed{} \times \boxed{} \times \boxed{} = \boxed{}$

Summary

Write 2–3 sentences to summarize the lesson.

Rate Yourself!

How confident are you about multiplying rational numbers? Check the box that applies.

For more help, go online to access a Personal Tutor. [Tutor]

FOLDABLES *Time to update your Foldable!*

Mid-Chapter Check

Vocabulary Check

1. **CCSS** **Be Precise** Define *rational number*. Give two examples of rational numbers. (Lesson 2)

_____

_____

_____

2. Explain how to write a fraction as a decimal. (Lesson 1)

_____

_____

Skills Check and Problem Solving

Fill in each ⬭ with <, >, or = to make a true sentence. (Lesson 1)

3. $\frac{3}{9}$ ⬭ $0.\overline{3}$

4. $-\frac{3}{8}$ ⬭ -0.5

5. $4.\overline{25}$ ⬭ $\frac{17}{4}$

6. **STEM** Africa makes up about $\frac{1}{5}$ of Earth's entire land area. Use the table to write the fraction of Earth's land area that is made up by the continents listed in the table. Write each in simplest form. (Lesson 2)

_____

_____

Continent	Decimal Portion of Earth's Land
Antarctica	0.095
Asia	0.295
Europe	0.07
North America	0.16

Find each product. Write in simplest form. (Lesson 3)

7. $\frac{5}{18} \cdot \frac{4}{15} =$ _____

8. $-1\frac{1}{2} \cdot \frac{2}{3} =$ _____

9. $2\frac{1}{3} \cdot 2\frac{1}{7} =$ _____

10. **Standardized Test Practice** Which expression is equal to $-\frac{2}{5}$ if $w = -3$, $x = \frac{3}{4}$, $y = -\frac{4}{5}$, and $z = -2\frac{2}{9}$? (Lesson 3)

Ⓐ $-wyz$

Ⓒ $\frac{2}{3}xy$

Ⓑ $5xz$

Ⓓ $-\frac{1}{2}wx$

A Flair for Fashion!

Use the information in the table to solve each problem. Write in simplest form.

1. For a size 8, does Dress A or Dress B require more fabric? Explain. _____

   _____

2. Write the amount of fabric needed to make Dress B in a size 12 as a decimal. _____

3. Estimate how many yards of fabric are needed to make ten of Dress A in a size 14. Then find the actual amount of fabric.

   _____

   _____

4. How much fabric is needed to make twelve of Dress A in a size 8? _____

5. A designer has half the amount of fabric needed to make Dress A in a size 10. How much fabric does she have? _____

6. A designer made a dress in a size 6 that uses $\frac{3}{4}$ the amount of fabric needed to make Dress B in a size 14. How much fabric does this new dress use? _____

Amount of Fabric Needed (yd)				
Dress	Size 8	Size 10	Size 12	Size 14
A	$3\frac{3}{8}$	$3\frac{1}{2}$	$3\frac{3}{4}$	$3\frac{7}{8}$
B	$3\frac{1}{4}$	$3\frac{1}{2}$	$3\frac{7}{8}$	4

Career Project

It's time to update your career portfolio! Use blogs and webpages of fashion designers to answer some of these questions: Where did they go to school? What was their first job? What do they say is the most difficult part about being a fashion designer? What inspires them to create their designs? What advice do they have for new designers?

_____

_____

_____

Suppose you are an employer hiring a fashion designer. What question would you ask a potential employee?

- _____

- _____

  _____

Dividing Rational Numbers

Getting Started

Scan Lesson 3-4 in your textbook. List two headings you would use to make an outline of the lesson.

- _____
  _____
  _____

- _____
  _____

Real-World Link

Global Literacy After learning the history of Mexico's holiday *El Día de los Muertos*, or *Day of the Dead*, students created a clay container to commemorate a loved one. They made their containers from two slabs of clay that that they cut into thirds. Complete the steps below to find the number of pieces used to make each container.

Step 1 Draw two rectangles to represent the two slabs of clay.

Step 2 On the same model, show each slab cut into thirds.

1. How many pieces are used for each container? ☐

   So, $2 \div \frac{1}{3} = $ ☐ .

2. Draw a diagram to find $3 \div \frac{1}{4}$.

   So, $3 \div \frac{1}{4} = $ ☐ .

3. How are $3 \div \frac{1}{4}$ and 3×4 related? _____
   _____

Notes

Divide Fractions

Fill in the boxes to find each quotient. Write in simplest form.

1. $\dfrac{2}{3} \div \dfrac{7}{3} = \dfrac{\boxed{}}{\boxed{}} \cdot \dfrac{\boxed{}}{\boxed{}} = \dfrac{\boxed{}}{\boxed{}}$

2. $\dfrac{3}{4} \div \dfrac{5}{8} = \dfrac{\boxed{}}{\boxed{}} \cdot \dfrac{\boxed{}}{\boxed{}} = \dfrac{\boxed{}}{\boxed{}}$ or $\boxed{}\dfrac{\boxed{}}{\boxed{}}$

3. Explain how to divide a fraction by a fraction.

_____

_____

Divide Algebraic Expressions

Simplify each expression.

4. $\dfrac{x^2}{4} \div \dfrac{xy}{2} = \boxed{}$

5. $\dfrac{b}{6ab} \div \dfrac{3b}{a} = \boxed{}$

6. $\dfrac{7}{gh} \div \dfrac{5}{4fh} = \boxed{}$

7. $\dfrac{14x}{xy} \div \dfrac{1}{10xy} = \boxed{}$

8. $\dfrac{q}{12} \div \dfrac{n^2}{2} = \boxed{}$

9. $\dfrac{b}{2d} \div \dfrac{2}{9c} = \boxed{}$

Summary

Write 2–3 sentences to summarize the lesson.

_____

_____

_____

_____

_____

_____

_____

Rate Yourself!

Are you ready to move on? Shade the section that applies.

I have a few questions.

I'm ready to move on.

I have a lot of questions.

For more help, go online to access a Personal Tutor.

FOLDABLES Time to update your Foldable!

Adding and Subtracting Like Fractions

Getting Started

Scan Lesson 3-5 in your textbook. Predict two things you will learn about adding and subtracting like fractions.

- _____
- _____

Real-World Link

Technology In a survey, users of e-Readers were asked to describe why they prefer e-Readers over books. The two most common responses are shown below.

Reason	Fraction of Responses
It's possible to change the font size and read faster.	$\frac{1}{8}$
The device is portable and convenient.	$\frac{5}{8}$

Find the fraction of people surveyed that chose the two most common responses.

Step 1 Divide the vertical number line shown into eighths.

Step 2 Graph a point that shows the fraction of people in the survey who chose the first reason.

Step 3 From that point, add the fraction of people who chose the second reason.

Step 4 Graph a second point to show the sum.

So, $\frac{1}{8} + \frac{5}{8} = \boxed{}$ or $\boxed{}$.

What fraction of people surveyed chose the reasons listed in the table? _____

Notes

Add Like Fractions

Use each model to find the sum. Write in simplest form.

1. $\frac{5}{12} + \frac{3}{12}$

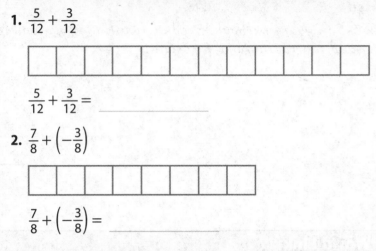

$\frac{5}{12} + \frac{3}{12} = $ _____

2. $\frac{7}{8} + \left(-\frac{3}{8}\right)$

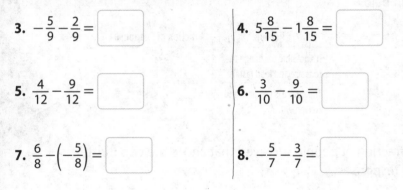

$\frac{7}{8} + \left(-\frac{3}{8}\right) = $ _____

Subtract Like Fractions

Find each difference. Write in simplest form.

3. $-\frac{5}{9} - \frac{2}{9} = $ ⬚

4. $5\frac{8}{15} - 1\frac{8}{15} = $ ⬚

5. $\frac{4}{12} - \frac{9}{12} = $ ⬚

6. $\frac{3}{10} - \frac{9}{10} = $ ⬚

7. $\frac{6}{8} - \left(-\frac{5}{8}\right) = $ ⬚

8. $-\frac{5}{7} - \frac{3}{7} = $ ⬚

Summary

Write 2–3 sentences to summarize the lesson.

_____

_____

_____

_____

_____

_____

_____

Rate Yourself!

Are you ready to move on? Shade the section that applies.

YES   ?   NO

For more help, go online to access a Personal Tutor.

Tutor

FOLDABLES Time to update your Foldable!

Lesson 3-6

Adding and Subtracting Unlike Fractions

Getting Started

Scan Lesson 3-6 in your textbook. List two real-world scenarios in which you would solve a problem by adding or subtracting fractions with different denominators.

- _____
  _____
  _____

- _____
  _____
  _____

Real-World Link

Climate The Sahara Desert is one of the driest places in the world, receiving only a couple inches of rainfall each year! The total amount of rainfall for three months is shown in the table.

Month	Rainfall (in.)
January	$\frac{3}{4}$
April	$\frac{1}{8}$
November	$\frac{1}{2}$

1. What is the least common multiple of 4, 8, and 2? ☐

2. Write equivalent fractions for $\frac{3}{4}$ and $\frac{1}{2}$ that have denominators equal to the least common multiple.

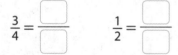

$$\frac{3}{4} = \frac{\square}{\square} \qquad \frac{1}{2} = \frac{\square}{\square}$$

3. Use the number line to find the total rainfall in the Sahara Desert for January, April, and November.

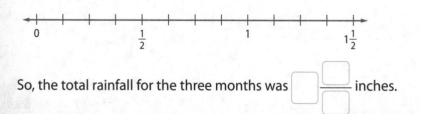

So, the total rainfall for the three months was ☐ $\frac{\square}{\square}$ inches.

Notes

Add Unlike Fractions

Fill in the boxes to find the sum.

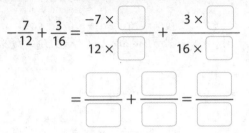

$$-\frac{7}{12} + \frac{3}{16} = \frac{-7 \times \boxed{}}{12 \times \boxed{}} + \frac{3 \times \boxed{}}{16 \times \boxed{}}$$

$$= \frac{\boxed{}}{\boxed{}} + \frac{\boxed{}}{\boxed{}} = \frac{\boxed{}}{\boxed{}}$$

Subtract Unlike Fractions

Complete the graphic organizer to find $\frac{5}{6} - \frac{3}{8}$.

Step 1	Find the _____ .	_____
Step 2	Rename the fractions using the _____ .	
Step 3	Subtract the fractions.	$\dfrac{\boxed{}}{\boxed{}} - \dfrac{\boxed{}}{\boxed{}} = \dfrac{\boxed{}}{\boxed{}}$

Summary

Write 2–3 sentences to summarize the lesson.

_____

_____

_____

_____

_____

_____

_____

Rate Yourself!

How confident are you about adding and subtracting unlike fractions? Shade the ring on the target.

For more help, go online to access a Personal Tutor.

FOLDABLES *Time to update your Foldable!*

Chapter Review

Vocabulary Check

Reconstruct the vocabulary word and definition from the letters under the grid. The letters for each column are scrambled directly under that column.

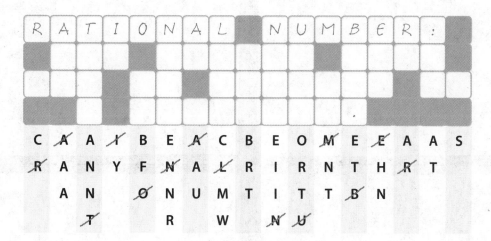

Complete each sentence using the vocabulary list at the beginning of the chapter.

1. The number 4.7 is a _____ decimal.

2. A _____ is another name for the multiplicative inverse.

3. The fractions $\frac{4}{6}$ and $\frac{1}{3}$ are _____ fractions.

4. _____ fractions are fractions that have the same denominator.

5. _____ decimals use a bar to show which digits repeat.

6. When a bar or line is placed over repeating digits, the number is written using _____.

7. The product of a number and its _____ is 1.

Use Your FOLDABLES

Use your Foldable to help review the chapter.

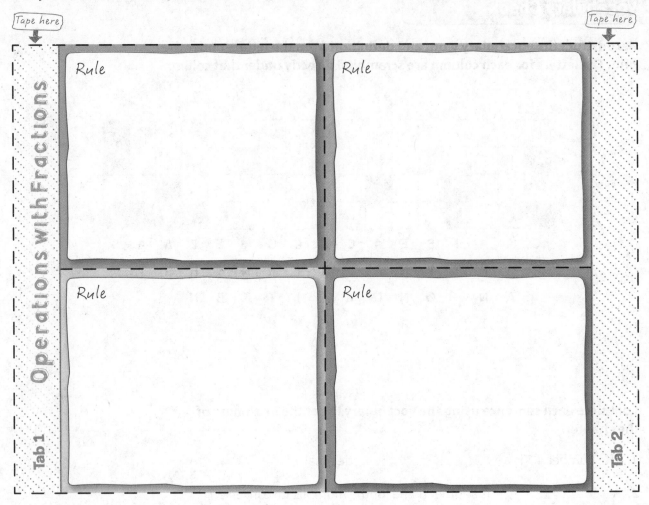

Got it?

Match each expression with its solution.

1. $\frac{5}{8} \cdot \frac{2}{3}$

2. $\frac{4}{5} \div \frac{2}{5}$

3. $\frac{8}{15} - \frac{4}{15}$

4. $\frac{1}{5} + \frac{2}{7}$

5. $3\frac{1}{4} + 2\frac{1}{2}$

a. $\frac{17}{35}$

b. $5\frac{3}{4}$

c. 2

d. $\frac{5}{12}$

e. $\frac{4}{15}$

Problem Solving

1. A regular incandescent light bulb can have a life of 750 hours while a compact fluorescent bulb (CFB) can have a life of 10,000 hours. What part of the life of a CFB is the life of an incandescent bulb? Write as a decimal. (Lesson 1)

Show your work.

2. A player's batting average equals the number of hits divided by the number of times at bat. A player has a batting average of .225, or 0.225. Write the player's batting average as a fraction in simplest form. (Lesson 2)

3. In a recent survey, $\frac{2}{3}$ of the people surveyed said they listen to Top 40 radio stations. Of these, $\frac{3}{4}$ said they listen to station WABC. What fraction of the people surveyed listen to station WABC? Write in simplest form and justify your answer. (Lesson 3)

4. A gallon of milk contains 128 ounces. Jung's favorite souvenir glass holds $10\frac{1}{2}$ ounces of milk. About how many of these glasses of milk can Jung pour from one gallon? (Lesson 4)

5. Mrs. Hamre wants to mail the packages shown. What is the total weight of the packages? Write in simplest form. (Lesson 5)

Package	Weight (oz)
1	$8\frac{7}{8}$
2	$5\frac{3}{8}$

6. Members of the Drama Club are building a set for a play. Rob needs a piece of wood $7\frac{3}{4}$ feet long for the front and a piece $4\frac{5}{8}$ feet long for the side. The wood comes in 12-foot lengths. After he cuts off the piece for the front, will Rob have enough left for the side? Explain your reasoning. (Lesson 6)

Reflect

 Answering the Essential Question

Use what you learned about operations with rational numbers to complete the graphic organizer. Describe a process to perform each operation.

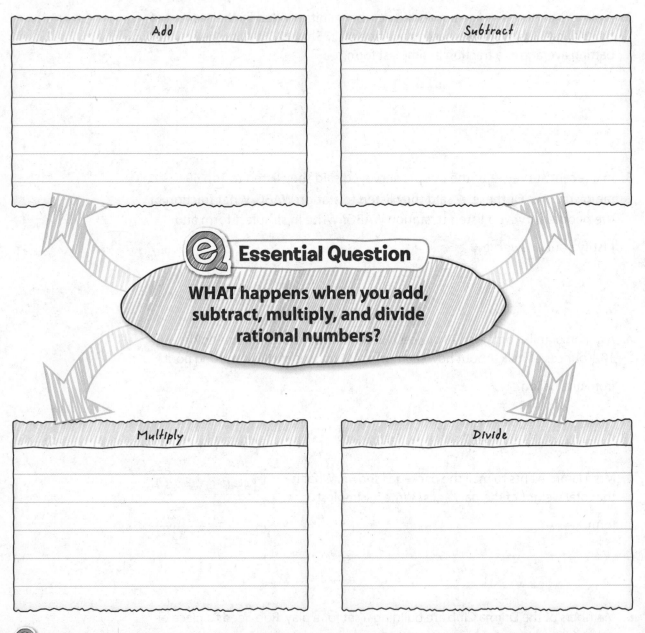

Add

Subtract

Essential Question

WHAT happens when you add, subtract, multiply, and divide rational numbers?

Multiply

Divide

Answer the Essential Question. WHAT happens when you add, subtract, multiply, and divide rational numbers?

Chapter 4

Powers and Roots

Vocabulary

base	monomial	power	square root
cube root	negative exponent	radical sign	standard form
exponent	perfect cube	real number	
irrational number	perfect square	scientific notation	

Vocabulary Activity

Complete the graphic organizer.

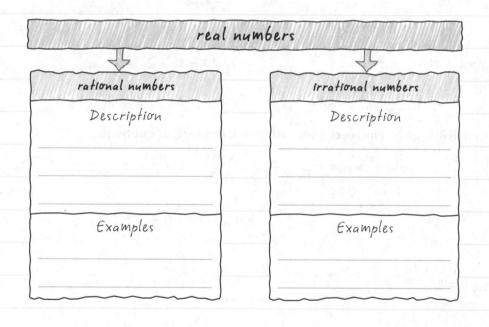

Are You Ready?

Try the Quick Check below.
Or, take the Online Readiness Quiz. Check ✓

Example 1

Evaluate $\frac{xy}{z}$ if $x = 3$, $y = -8$, and $z = 6$.

$\frac{xy}{z} = \frac{3(-8)}{6}$ Replace x with 3, y with −8, and z with 6.

$\quad = \frac{-24}{6}$ Multiply.

$\quad = -4$ Divide.

Example 2

Find $-25 - (-36)$.

$-25 - (-36)$

$\quad = -25 + 36$ To subtract −36, add 36.

$\quad = 11$ Simplify.

Quick Check

Expressions Evaluate each expression if $a = 4$, $b = -7$, and $c = 5$.

1. $3a + 2c =$ _____

Show your work.

2. $\frac{ac}{2} =$ _____

3. $-5b + 6a =$ _____

4. $\frac{3}{4}(2bc) =$ _____

5. $\frac{ab}{-4} + 2b =$ _____

6. $4b - 3c + 2a =$ _____

Operations with Integers Find each sum, difference, product, or quotient.

7. $-27 + (-13) =$ _____

8. $15 - (-20) =$ _____

9. $6(-10) =$ _____

10. $-25(-4) =$ _____

11. $-36 \div (-9) =$ _____

12. $-54 \div 6 =$ _____

How Did You Do?

Which problems did you answer correctly in the Quick Check?
Shade those exercise numbers below.

① ② ③ ④ ⑤ ⑥ ⑦ ⑧ ⑨ ⑩ ⑪ ⑫

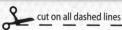 cut on all dashed lines   fold on all solid lines   tape to page 90   FOLDABLES

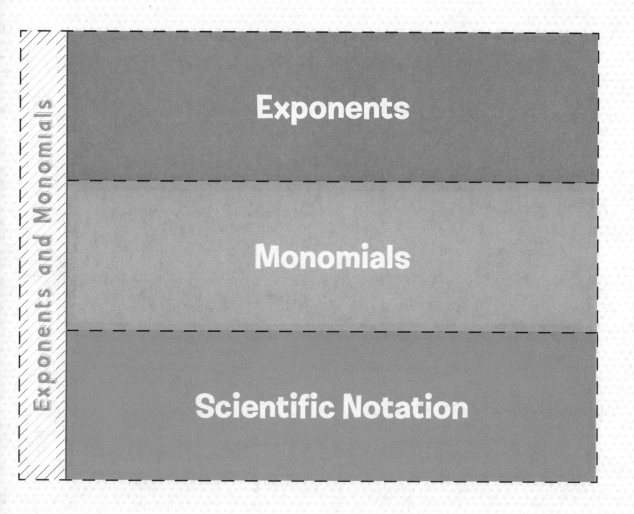

Exponents and Monomials

Exponents

Monomials

Scientific Notation

1 Cut out the Foldable above.

2 Place your Foldable on page 90.

3 Use the Foldable throughout this chapter to help you learn about exponents and monomials.

Examples

Examples

Examples

Powers and Exponents

Getting Started

Scan Lesson 4-1 in your textbook. List two headings you would use to make an outline of the lesson.

- _____
  _____

- _____
  _____

Vocabulary Start-Up

A **power** is a number that is expressed using an exponent, such as 2^4. In 2^4, the **base** is 2, and the **exponent** is 4. This means that 2 is used as a factor 4 times. So, $2^4 = 2 \times 2 \times 2 \times 2$, or 16.

Label the expression with the terms *base*, *exponent*, and *power*.

_____ { 5^2 ——— _____

_____

For each power, underline the base and circle the exponent.

$$5^2 \qquad 6^3 \qquad 10^4 \qquad 7^8$$

Real-World Link

Computers Data storage capacity is measured in bytes and is based on powers of 2. The standard scientific meanings for the prefixes *mega-* and *giga-* are one million and one billion, respectively.

1. Write 2^{20} in expanded form. Then find its value. _____

_____

2. Why do you think the term *megabyte* is used to represent 2^{20} bytes?

_____

_____

3. What is the approximate value of one gigabyte, or 2^{30} bytes?

_____

Notes

Use Exponents

Match each verbal expression with its numerical expression. Choices may be used more than once.

1. eight to the second power

2. eight cubed

3. eight to the fourth power

4. eight squared

a. 8^4

b. 8^3

c. 8^2

d. 2^8

Evaluate Expressions

Complete the graphic organizer by evaluating the algebraic expression for each set of values.

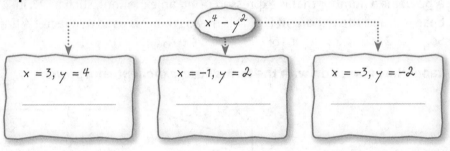

$x^4 - y^2$

x = 3, y = 4 ____

x = -1, y = 2 ____

x = -3, y = -2 ____

Suppose a negative number is raised to a power. When is the result positive?

____

____

Summary

Write 2–3 sentences to summarize the lesson.

____

____

____

____

____

____

____

____

____

Rate Yourself!

How confident are you about using exponents? Shade the ring on the target.

I'm on target.

I need help.

For more help, go online to access a Personal Tutor.

FOLDABLES Time to update your Foldable!

Lesson 4-2
Negative Exponents

Getting Started

Scan Lesson 4-2 in your textbook. Predict two things you will learn about negative exponents.

* _____

  _____

  _____

* _____

  _____

Quick Review

Simplify the expression below. Use an exponent.

$$\frac{2 \cdot 2 \cdot 2}{2 \cdot 2 \cdot 2 \cdot 2 \cdot 2} = \boxed{}$$

🌎 Real-World Link

Snowflakes Have you ever heard that no two snowflakes are exactly alike? This is because they are made up of water molecules, which grow at varying patterns and rates. The table shows the diameters of three snowflakes.

Snowflake	Diameter (cm)
A	0.01
B	0.1
C	1

1. How many times as great is the diameter of snowflake C than the diameter of snowflake B? _____

2. How many times as great is the diameter of snowflake C than the diameter of snowflake A? _____

   Write this number using an exponent. $\boxed{}$

3. The diameter of snowflake B is 0.1 times as great as the diameter of snowflake C. Write this number as a fraction. $\boxed{}$

4. The diameter of snowflake A is how many times as great as the diameter of snowflake C? _____

   Write this number as a fraction. $\boxed{}$

   Write the fraction with an exponent in the denominator. $\boxed{}$

5. Complete the table.

Number	1000	100	10	1	$\frac{1}{10}$	$\frac{1}{100}$	$\frac{1}{1000}$
Power	10^3	10^2	10^1	10^0	10^{-1}		

   Predict how to write $\frac{1}{x^4}$ using a negative exponent. $\boxed{}$

Notes

Negative Exponents

Circle the correct answer.

1. $4^{-3} = -64$ or $\dfrac{1}{64}$

2. $6^{-2} = \dfrac{1}{36}$ or -36

3. $\dfrac{1}{25} = 5^{-2}$ or $\dfrac{1}{5^{-2}}$

4. $\dfrac{1}{216} = \dfrac{1}{6^{-3}}$ or 6^{-3}

5. $3^{-4} = -81$ or $\dfrac{1}{81}$

6. $\dfrac{1}{32} = 2^{-5}$ or $\dfrac{1}{2^{-5}}$

7. Fill in the boxes to write 0.001 in three other ways.

Decimal		Fraction		Expression with Positive Exponent		Expression with Negative Exponent
0.001	=		=		=	

Evaluating Expressions with Negative Exponents

If $a = -2$ and $b = 4$, match each expression with its correct value.

8. a^{-3}

9. 6^a

10. b^a

11. -3^b

a. -81

b. $\dfrac{1}{16}$

c. $-\dfrac{1}{8}$

d. $\dfrac{1}{36}$

Summary

Write 2–3 sentences to summarize the lesson.

Rate Yourself!

Are you ready to move on? Shade the section that applies.

I have a few questions. | I'm ready to move on.

I have a lot of questions.

For more help, go online to access a Personal Tutor.

Tutor

FOLDABLES Time to update your Foldable!

Multiplying and Dividing Monomials

Getting Started

Scan Lesson 4-3 in your textbook. Describe the Product of Powers Property and the Quotient of Powers Property using your own words.

- Product of Powers: _____

  _____

- Quotient of Powers: _____

  _____

🌐 Real-World Link

Football The playing field in football is 120 yards long and $53\frac{1}{3}$ yards wide. Recently, a local high school cleared a rectangular area that measured 2^7 yards long and 2^6 yards wide to make room for a new football field.

1. How would you find the area of the space that was cleared for the new field?

   _____

2. What are the length and width of the space?

   length: _____   width: _____

3. Write an expression that could be used to find the area of the space for the new field. _____

4. What is the expression from Exercise 3 written as a product of its factors?

   _____

   How could you rewrite this using exponents? _____

5. How is the exponent from Exercise 4 related to the exponents from Exercise 3? _____

   _____

6. Is the area that was cleared large enough for a regulation football field? Explain your reasoning.

   _____

   _____

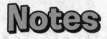

Notes

Multiply Monomials

Fill in the blanks to find each product.

1. $4^3 \cdot 4^2 = 4^{\boxed{}+\boxed{}} = 4^{\boxed{}}$

2. $2^5 \cdot 2^3 = 2^{\boxed{}+\boxed{}} = 2^{\boxed{}}$

3. $5^{-3} \cdot 5^6 = 5^{\boxed{}+\boxed{}} = 5^{\boxed{}}$

4. $2y^2 \cdot (-3y^{-5}) = \boxed{}\, y^{\boxed{}+\boxed{}} = \boxed{}\,\boxed{}^{\boxed{}}$

5. $5x^{-6} \cdot 3x^{-2} = \boxed{}\,\boxed{}^{\boxed{}+\boxed{}} = \boxed{}\,\boxed{}^{\boxed{}}$

Divide Monomials

Cross out the one that does not belong. Then state the relationship among the three remaining expressions in the circle.

_____

_____

_____

$\dfrac{7^5}{7^2}$ $\dfrac{x^2}{x^{-1}}$

$\dfrac{-y^6}{-y^3}$ $\dfrac{-3^3}{-3}$

Summary

Write 2–3 sentences to summarize the lesson.

_____

_____

_____

_____

_____

_____

_____

Rate Yourself!

How confident are you about multiplying and dividing monomials? Check the box that applies.

For more help, go online to access a Personal Tutor.

FOLDABLES *Time to update your Foldable!*

Scientific Notation

Getting Started

Scan Lesson 4-4 in your textbook. List two-real-world scenarios in which you would use scientific notation.

- _____
  _____

- _____
  _____

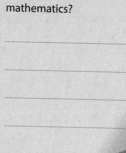

Vocabulary

How can the English definition of *standard* help you remember how *standard form* is used in mathematics?

_____

_____

_____

Real-World Link

Space Earth is the third planet from the Sun in our solar system. Because Earth's rotation about the Sun is not circular, the maximum distance between Earth and the Sun is about 95 million miles and the minimum distance is about 91 million miles.

1. Write 91 million using numbers. _____

2. What is the value of 9.1×10? 9.1×100? _____

3. Complete the table.

Expression	Product
9.1×10	
9.1×100	
9.1×1000	
$9.1 \times$ _____	
$9.1 \times$ _____	
$9.1 \times$ _____	
$9.1 \times$ _____	

4. Multiplying 9.1 by what number results in the original number?

_____

5. What is this number written as a power of 10? _____

6. How could you rewrite 91 million as the product of a number and

a power of 10? _____

7. Based on your answer to Exercise 6, what is 95 million written as the

product of a number and a power of 10? _____

Notes

Scientific Notation

Circle each number below that is written in scientific notation.

0.1×10^{-8}	2.019×10^{12}	5×10^{-22}
9.5×10^{-7}	10.6×10^{14}	0.526×10^{-2}
4.00001×10^9	8.496×10^8	6.0×10^{-10}

Compare and Order Numbers Written in Scientific Notation

Write each number in standard form. Then order each set of numbers from least to greatest.

1. 4.05×10^5, 420,000, 3×10^5, 1.3×10^5

2. 0.0024, 2×10^{-2}, 3.1×10^3, 0.029

3. How do you compare numbers in scientific notation?

Summary

Write 2–3 sentences to summarize the lesson.

Rate Yourself!

☐ *I understand how to write numbers in scientific notation.*

▶▶ Great! You're ready to move on!

☐ *I still have questions about writing numbers in scientific notation.*

📖 No Problem! Go online to access a Personal Tutor.

 **FOLDABLES** *Time to update your Foldable!*

Mid-Chapter Check

Vocabulary Check

1. **CCSS Be Precise** Define *power* using the words *base* and *exponent*. Give an example of a power and describe the base and exponent. (Lesson 1)

_____

_____

_____

2. How do you multiply monomials with the same base? (Lesson 3)

_____

_____

Skills Check and Problem Solving

3. **STEM** The wavelength of a microwave is 10^{-2} meter. Write the wavelength of a microwave as a fraction without a negative exponent.

(Lesson 2) _____

Find each product or quotient. Express using exponents. (Lesson 3)

4. $8^4 \cdot 8^5 =$ _____

5. $\dfrac{6^7}{6^2} =$ _____

6. $\dfrac{x^7}{x^2} =$ _____

7. $2y^3 \cdot 7y^3 =$ _____

Write each number in scientific notation. (Lesson 4)

8. $6{,}492{,}000 =$ _____

9. $0.00951 =$ _____

10. **Standardized Test Practice** The number of acres consumed by a forest fire triples every two hours. Which of the following expressions represents the number of acres consumed after 1 day? (Lesson 1)

Hours	2	4	6	8
Acres Consumed	3^1	3^2	3^3	3^4

Ⓐ 3^{10} acres

Ⓑ 3^{12} acres

Ⓒ 3^{18} acres

Ⓓ 3^{24} acres

Relying on Robots

Use the information in the table to solve each problem.

1. Write the mass of the robot in standard form.

   _____

2. Write the length of the robot in scientific

   notation. _____

3. Write the leg diameter of the robot in scientific

   notation. _____

4. What is the mass in milligrams? Write in standard form. (*Hint*: 1 kg = 1,000,000 mg)

   _____

5. Real insects called water striders can travel 8.3 times faster than the robot. Write the speed of water striders in scientific notation. _____

Robotic Insect Characteristics	
Mass	3.5×10^{-4} kg
Length	0.09 m
Leg Diameter	0.2 mm
Speed	180 mm/s

Career Project

It's time to update your career portfolio! Investigate the education and training requirements for a career in robotics engineering.

_____

_____

_____

_____

_____

_____

What skills would you need to improve to succeed in this career?

- _____
- _____
- _____
- _____
- _____

Compute with Scientific Notation

Getting Started

Scan Lesson 4-5 in your textbook. List two headings you would use to make an outline of the lesson.

- _____
  _____
  _____
- _____
  _____
  _____

Real-World Link

Aircraft The SR-71 Blackbird is one of the world's fastest airplanes. It is capable of traveling at a cruising speed of Mach 3, or three times the speed of sound. The speed of sound is approximately 760 miles per hour.

1. Express 760 in scientific notation.

   _____

2. Suppose a SR-71 Blackbird traveled 150 hours in one month. Express 150 in scientific notation.

   _____

3. Write a multiplication expression using the numbers in Exercises 1 and 2 to represent the total number of miles traveled by the SR-71 Blackbird in one month.

   _____

4. You can use the Commutative and Associative Properties of Multiplication to rewrite the expression in Exercise 3 as $(7.6 \times 1.5) \times (10^2 \times 10^2)$. Fill in the boxes to write an equivalent expression with one power of 10.

   ☐ $\times 10^{☐}$

5. Write the product in Exercise 4 in both scientific notation and standard form.

   _____

Notes

Multiplication and Division with Scientific Notation

For each product, fill in the boxes to show how you would apply the Commutative and Associative Properties.

1. $(1.5 \times 10^5)(6.2 \times 10^3)$

$$\left(\boxed{} \times \boxed{} \right) \left(\boxed{} \times \boxed{} \right)$$

2. $\dfrac{1.6 \times 10^6}{3.2 \times 10^4}$

$$\left(\boxed{} \div \boxed{} \right) \left(\boxed{} \div \boxed{} \right)$$

3. How do you divide two numbers written in scientific notation?

_____

_____

Addition and Subtraction with Scientific Notation

For each expression, circle the number that you would rewrite in order to add or subtract. Then rewrite the number so that both numbers have the same exponent.

4. $(8.93 \times 10^5) + (7.12 \times 10^4)$ _____

5. $(6.142 \times 10^5) - 12,430,000$ _____

6. $529,000 + (3.29 \times 10^6)$ _____

Summary

Write 2–3 sentences to summarize the lesson.

_____

_____

_____

_____

_____

_____

_____

_____

Rate Yourself!

How well do you understand computing with scientific notation? Circle the image that applies.

Clear   Somewhat Clear   Not So Clear

For more help, go online to access a Personal Tutor.

Square Root and Cube Roots

Getting Started

Scan Lesson 4-6 in your textbook. Write the definitions of square root and cube root.

- square root _____

_____

- cube root _____

_____

Real-World Link

Rainforest Tropical rainforests contain the greatest diversity of plants and animals on Earth—and they cover less than 5 percent of Earth's land! Just a four square mile patch of rainforest contains thousands of species of plants and trees, and hundreds of species of mammals, birds, reptiles, and amphibians.

1. Two integer factors of 4 are 1 and 4. Write three other pairs of integers that have products of 4.

pair 1: ☐ and ☐

pair 2: ☐ and ☐

pair 3: ☐ and ☐

2. Which pair(s) of integers that you wrote for Exercise 1 could represent the dimensions of an area of land? _____

3. Which pair of integers could represent the dimensions of a square area of land? _____

4. What are the dimensions of a four square mile patch of land?

_____

5. What would be the dimensions of the following square patches of land?

a nine square mile patch of land _____

a twenty five square kilometer patch of land _____

Notes

Find Square Root

Match each perfect square with its square roots.

1. 1.69

2. 1.44

3. 2.25

4. 1.96

a. ±1.1

b. ±1.2

c. ±1.3

d. ±1.4

e. ±1.5

5. Why does a perfect square have two square roots?

_____

_____

_____

Find Cube Root

For each pair of numbers, circle the perfect cube. Then write the cube root of the circled number on the line provided.

6. 8   4   _____

7. 25   125   _____

8. 0.16   −0.64   _____

9. $\frac{125}{216}$ $\frac{25}{36}$ _____

Summary

Write 2–3 sentences to summarize the lesson.

_____

_____

_____

_____

_____

_____

_____

Rate Yourself!

Are you ready to move on? Shade the section that applies.

I have a few questions.

I'm ready to move on.

I have a lot of questions.

For more help, go online to access a Personal Tutor.

The Real Number System

Getting Started

Scan Lesson 4-7 in your textbook. Predict two things you will learn about real numbers.

- _____
  _____
  _____

- _____
  _____
  _____

Vocabulary Start-Up

You have learned that rational numbers, such as $4\frac{1}{2}$, 0.15, and $\sqrt{25}$, are numbers that can be written as fractions. **Irrational numbers**, such as π and $\sqrt{15}$, are numbers that cannot be written as fractions.

Label the diagram with the terms *whole*, *integer*, *rational*, and *irrational*. Then complete the diagram using the numbers from the number bank.

Real Numbers

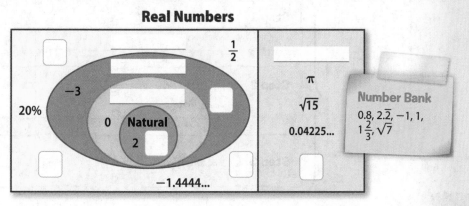

$\frac{1}{2}$

π

$\sqrt{15}$

0.04225...

−3

20%

0   Natural

2

−1.4444...

Number Bank

$0.8, 2.\overline{2}, -1, 1,$
$1\frac{2}{3}, \sqrt{7}$

Real-World Link

Weather Meteorologists use the formula $t^2 = \frac{d^3}{216}$ to predict the time t in hours a thunderstorm will last when it is d miles across.

1. Suppose a thunderstorm is 6 miles across. Write and solve an equation to determine how long the thunderstorm will last. _____

2. Suppose a thunderstorm is 12 miles across. About how long will the thunderstorm last? Round to the nearest whole number. _____

Notes

Identify and Compare Real Numbers

Circle the correct phrase to complete each sentence. Explain your reasoning.

1. $0.6\overline{7}$ is (less than, greater than) 0.67.

_____

2. $\sqrt{33}$ is (less than, greater than) 6.

_____

3. 2.5 is (less than, greater than) $\sqrt{8}$.

_____

4. $\sqrt[3]{100}$ is (less than, greater than) 400%.

_____

Solve Equations

Complete the organizer to solve the equation $x^2 = 10$.

Solve equations with square roots.

Step 1	Write the equation.	⋯⋯▶	
Step 2	Definition of Square Root	⋯⋯▶	
Step 3	Use a calculator.	⋯⋯▶	

Summary

Write 2–3 sentences to summarize the lesson.

_____

_____

_____

_____

_____

_____

Rate Yourself!

Are you ready to move on? Shade the section that applies.

YES   ?   NO

For more help, go online to access a Personal Tutor.

Tutor

Chapter Review

Vocabulary Check

Fill in the blank with the correct vocabulary term. Then circle the word that completes the sentence in the word search.

1. The number 5.2×10^4 is written in _____ notation.

2. The _____ tells how many times a number is used as a factor.

3. A _____ root of a number is one of its three equal factors.

4. In 5^7, the number 5 is the _____.

5. The square of an integer is a _____ square.

6. The set of _____ numbers is made up of the set of rational numbers and the set of irrational numbers.

7. A _____ sign is used to indicate a nonnegative square root.

8. A _____ root of a number is one of its two equal factors.

9. A number is written in _____ form when it does not contain exponents.

10. A(n) _____ is a number, a variable, or a product of a number and one or more variables.

11. The number 6^4 is a _____.

12. A decimal that does not repeat or terminate is a(n) _____ number.

R	J	D	F	X	D	V	S	P	O	W	E	R	H	K	S	L	C
T	M	J	Y	C	B	T	Y	P	Q	C	E	W	G	M	L	A	M
V	U	S	F	A	A	E	T	D	S	A	A	T	Q	U	I	C	J
G	R	S	L	N	S	W	V	O	L	D	T	P	S	R	S	I	F
I	D	R	D	A	E	H	Y	M	O	H	O	C	P	H	O	D	O
X	R	A	D	A	D	L	W	W	S	S	D	X	E	H	B	A	C
N	R	R	K	F	L	H	T	I	W	D	G	R	A	F	M	R	D
D	W	R	A	V	M	Y	I	M	X	T	A	Y	E	P	R	I	L
M	Z	T	U	T	Z	G	Q	C	K	U	C	C	X	D	L	E	W
S	J	M	S	O	I	M	U	C	Q	M	C	M	P	E	D	Y	P
Z	R	Z	Y	V	O	O	C	S	U	Y	H	G	O	D	I	I	C
C	E	U	W	P	J	E	N	E	V	B	L	Y	N	P	O	L	P
L	A	I	M	O	N	O	M	A	F	U	E	R	E	G	V	K	Z
E	Y	S	N	N	J	I	G	X	L	D	M	H	N	V	D	I	L
S	C	I	E	N	T	I	F	I	C	L	A	Z	T	V	B	I	L
H	M	C	H	T	K	N	P	K	C	R	W	G	M	Z	A	E	W

Use Your FOLDABLES

Use your Foldable to help review the chapter.

Tape here

Exponents and Monomials

Description

Description

Description

Got it?

Circle the correct term or number to complete each sentence.

1. 5^{-2} is equal to $\left(-25, \frac{1}{25}\right)$.

2. A number that is written as a product of a power of 10 and a factor greater than or equal to 1 and less than 10 is in (scientific notation, standard form).

3. You would use the (Product of Powers, Quotient of Powers) rule to simplify the expression $a^5 \cdot a^3$.

4. Another way to write $\frac{x^8}{x^4}$ is (x^2, x^4).

5. $\sqrt{36}$ is equal to $(6, -6)$.

Problem Solving

1. A popular beverage company has about $(2^3)(3)(5^4)$ locations worldwide. How many locations do they have? Write the number in standard form. (Lesson 1)

_____

2. A dime has a thickness of approximately 0.001 meter. Write this decimal as a fraction and as a power of ten. (Lesson 2) _____

3. The table compares the number of people who drove to work to the number of people who walked to work. How many times more people drove to work than walked? (Lesson 3) _____

Mode of Transportation	Number of People
Drove	10^5
Walked	10^3

4. The length of an infrared light wave is approximately 0.0000037 meter. Write this number in scientific notation. (Lesson 4) _____

5. Neurons are cells in the nervous system that process and transmit information. An average neuron is about 5×10^{-6} meter in diameter. A standard tennis ball is 0.67 meter in diameter. About how many times as great is the diameter of a tennis ball than a neuron? (Lesson 5) _____

6. The formula $d = \sqrt{\frac{h}{0.57}}$, where h represents the number of feet above sea level, can be used to find the distance d in miles to the horizon. Ella is standing on a deck of a cruise ship and is approximately 80 feet above sea level. Estimate how far out to the horizon she can see. (Lesson 6) _____

7. The area of a cross section of Earth at the equator is approximately 127,796,483 square kilometers. Use the formula $A = 3.14r^2$ to find the approximate radius of Earth to the nearest kilometer. (Lesson 7) _____

Reflect

Answering the Essential Question

Use what you learned about numbers to complete the graphic organizer. For each category, describe why you would use that form for the number $35,036 \frac{1}{3}$. Then write the number in that form. If you would not use the number in that form, explain why.

Decimal

Power

Essential Question

WHY is it useful to write numbers in different ways?

Fraction

Scientific Notation

Answer the Essential Question. WHY is it useful to write numbers in different ways?

Chapter 5
Ratio, Proportion, and Similar Figures

Vocabulary

complex fraction	indirect measurement	scale
congruent	nonproportional	scale drawing
constant of proportionality	proportion	scale factor
corresponding parts	proportional	scale model
cross products	rate	similar figures
dimensional analysis	ratio	unit rate

Vocabulary Activity

Select four vocabulary terms from the list above. Write a definition for the terms based upon what you already know. As you go through the chapter, come back to this page and update your definitions if necessary.

Term	What I Know	What I Learned

CCSS Quick Review

Common Core Review 6.RP.1, 6.EE.7

Example 1

What fraction of the games played were wins? Write in simplest form.

Filmore Flyers	
Team Statistics	
Wins	15
Losses	10

total games: $15 + 10$ or 25 games

number of wins: 15

$$\frac{15}{25} = \frac{3}{5}$$

The Flyers won $\frac{3}{5}$ of their games.

Example 2

Solve $9.5x = 38$.

$9.5x = 38$	Write the equation.
$\dfrac{9.5x}{9.5} = \dfrac{38}{9.5}$	Division Property of Equality
$x = 4$	Simplify.

Check $\quad 9.5x \stackrel{?}{=} 38$

$\qquad 9.5(4) \stackrel{?}{=} 38$

$\qquad 38 = 38$ ✓

Quick Check

Fractions The table shows the number of students at Harrison Middle School that are enrolled in world language classes. Find the fraction of students that are enrolled in each language.

Number of Students Enrolled in World Language Classes	
Chinese	51
French	33
Spanish	42

1. French _____

2. Spanish _____

3. Chinese _____

Show your work. →

One-Step Equations Solve each equation. Check your solution.

4. $9m = 18$ _____

5. $42 = 7n$ _____

6. $4y = 10$ _____

7. $60x = 12$ _____

8. $7 = 3.5a$ _____

9. $16.5 = 5.5d$ _____

How Did You Do?

Which problems did you answer correctly in the Quick Check? Shade those exercise numbers below.

① ② ③ ④ ⑤ ⑥ ⑦ ⑧ ⑨

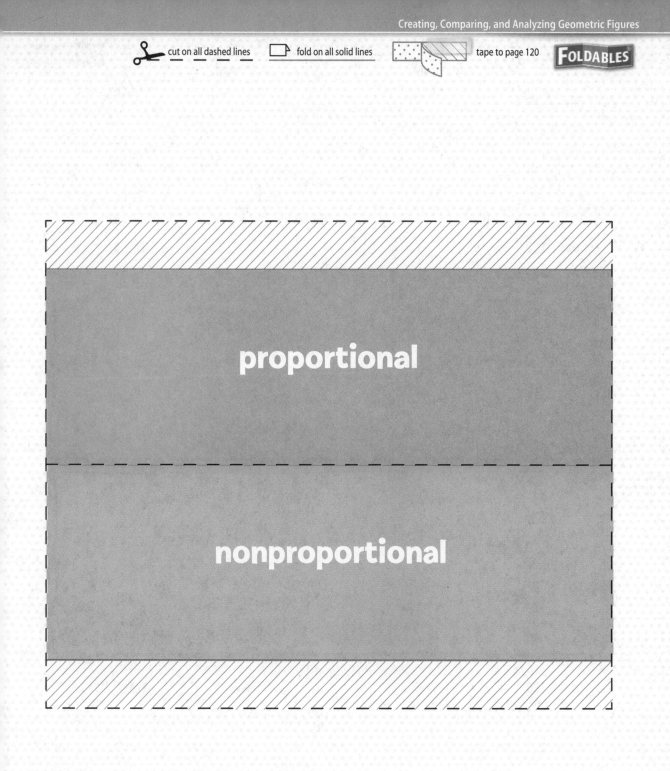

proportional

nonproportional

1 Cut out the Foldable above.

2 Place your Foldable on page 120.

3 Use the Foldable to help you learn about proportional relationships.

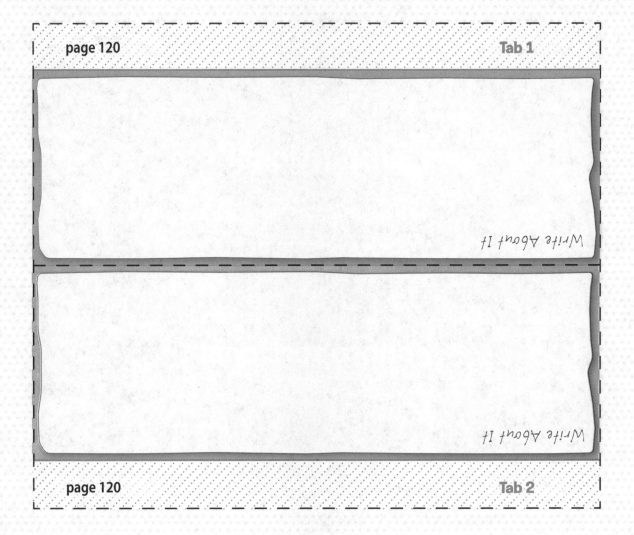

page 120

Tab 1

Write About It

Write About It

page 120

Tab 2

Getting Started

Scan Lesson 5-1 in your textbook. List two real-world scenarios in which you would use ratios.

- _____
- _____

Real-World Link

Surfing A surfer who rides inside the curl of a wave is said to be *tube riding*. The shape of the tube is described as *cylindrical, square, round,* or *almond,* depending on how its length compares to its width.

Value of $\frac{length}{width}$	Shape of Tube
less than 1	square
exactly 1	cylindrical
between 1 and 2	round
greater than 2	almond

1. The tube of a wave is 12 feet long and 15 feet wide. Write the fraction $\frac{length}{width}$ in simplest form. Then describe the shape of the tube.

2. The tube of a wave is 20 feet long and 12 feet wide. What is the shape of the tube? Explain.

_____

_____

3. A surfer is riding in a tube that is cylindrical. The width of the tube is 10 feet. What is the length? Explain how you know.

_____

_____

Notes

Ratios

Cross out the ratio that is not equivalent to *16 girls out of 24 students*.

$$\frac{2}{3} \qquad \frac{3}{4} \qquad \frac{8}{12} \qquad \frac{16}{24}$$

How did you decide which ratio to cross out?

_____

_____

_____

Simplify Ratios involving Measurements

Write each ratio as a fraction in simplest form.

1. 15 cans out of 9 cases $= \dfrac{\boxed{}\ \text{cans}}{\boxed{}\ \text{cases}} = \dfrac{\boxed{}}{\boxed{}}$

2. 16 inches to 4 feet $= \dfrac{\boxed{}\ \text{inches}}{\boxed{}\ \text{inches}} = \dfrac{\boxed{}}{\boxed{}}$

3. How is Exercise 2 different from Exercise 1?

_____

_____

_____

Summary

Write 2–3 sentences to summarize the lesson.

_____

_____

_____

_____

_____

_____

Rate Yourself!

Are you ready to move on? Shade the section that applies.

YES   ?   NO

For more help, go online to access a Personal Tutor.

Tutor

Unit Rates

Getting Started

Scan Lesson 5-2 in your textbook. List two headings you would use to make an outline of the lesson.

- _____
  _____
  _____

- _____
  _____

Quick Review

There are 15 girls to 13 boys in a class. Write the ratio of boys to girls.

_____

_____

Vocabulary Start-Up

A **rate** is a ratio of two measurements having different units. When a rate is simplified so that it has a denominator of 1, it is called a **unit rate**.

Complete the Venn diagram by writing the phrases below in the correct position.

$24 for 3 pounds
can be written as a fraction
15 students per teacher

$10 per hour
18 red cars to 6 blue cars

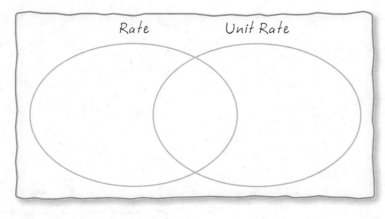

Rate   Unit Rate

Real-World Link

Texting When sending a text message to a friend, Markesha typed 90 characters in a little more than 60 seconds. If she typed at a constant rate, how many characters did she type each second?

_____

Notes

Find Unit Rates

Determine if each situation in the table represents a unit rate. Write Y for *yes* or N for *no*. If *no*, rewrite the situation as a unit rate.

Situation	Unit Rate Write Y or N	Changed Situation
60 pages read in 40 minutes		
50 miles per gallon		
$7 per movie ticket		
300 miles driven in 5 hours		
9 dollars for 2 sandwiches		
5 pencils for each student		

Compare Unit Rates

Fill in each ◯ with <, >, or = to compare the unit rates. Then explain why you chose the sign that you did.

1. 10 books for $12 ◯ 15 books for $18.75

_____

_____

2. 12 cans for $4.20 ◯ 20 cans for $6

_____

_____

Summary

Write 2–3 sentences to summarize the lesson.

_____

_____

_____

_____

_____

Rate Yourself!

How confident are you about unit rates? Check the box that applies.

☹ ☹ ☺

☐ ☐ ☐ ☐ ☐

For more help, go online to access a Personal Tutor.

Tutor

Complex Fractions and Unit Rates

Getting Started

Scan Lesson 5-3 in your textbook. Predict two things you will learn about complex fractions.

- _____

  _____

- _____

  _____

Quick Review

Find the value of $1\frac{1}{6} \div \frac{2}{3}$.

_____

Real-World Link

Music DJs use beats per minute, or BPM, to blend together songs that have similar tempos. Song A has 92 beats per minute. Song B has 23 beats per $\frac{1}{4}$ minute.

1. Fill in the boxes to write a ratio comparing the number of beats and the time in minutes for song A.

   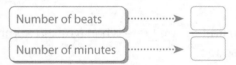

2. Fill in the boxes to write a ratio comparing the number of beats and the time in minutes for song B.

3. Is the ratio you wrote in Exercise 2 written in simplest form? Explain.

   _____

   _____

4. How could you simplify the ratio you wrote in Exercise 2?

   _____

5. Do the songs have the same number of beats per minute? Explain your reasoning.

   _____

   _____

Questions

Simplify Complex Fractions

Complete the organizer by following the steps to simplify $\frac{\frac{2}{3}}{4}$.

Step 1   Write as a division problem.

Step 2   Multiply by the reciprocal of $\frac{3}{4}$.

Step 3   Simplify.   .

Find Unit Rates

Express each rate as a unit rate.

1. driving 150 miles in $2\frac{1}{2}$ hours _____

2. flying 1700 miles in $3\frac{2}{5}$ hours _____

3. How do you find the unit rate when a mixed number is included in the complex fraction?

_____

_____

Summary

Write 2–3 sentences to summarize the lesson.

_____

_____

_____

_____

_____

_____

_____

Rate Yourself!

How confident are you about simplifying complex fractions? Shade the ring on the target.

For more help, go online to access a Personal Tutor.

Lesson 5-4
Converting Rates

Getting Started

Scan Lesson 5-4 in your textbook. List two real-world scenarios in which you would convert rates.

- _____
  _____

- _____
  _____

Vocabulary

Write the definition of *dimensional analysis* in your own words.

_____

_____

_____

Real-World Link

Sea Creature Leafy seadragons, which are members of the same family as seahorses, get their name from the leaf-like appendages that camouflage them. A leafy seadragon can travel about 14 feet in one hour.

You can use fractions involving units to calculate the average speed of the leafy seadragon in inches per minute.

1. How many inches are in 1 foot? 14 feet?

1 foot = _____ inches

14 feet = _____ inches

2. How many minutes are in 1 hour?

1 hour = _____ minutes

3. Use equivalent measurements to complete the following expression.

$$\frac{14\ ft}{1\ h} \times \frac{\boxed{}\ in.}{1\ ft} \times \frac{1\ h}{\boxed{}\ min}$$

4. Complete the following statement.

14 feet per hour = $\boxed{}$ inches per minute

5. How can you convert feet per hour to inches per minute?

_____

_____

Notes

Dimensional Analysis

Fill in the missing conversion factor. Then convert the unit rate.

1. $\dfrac{570 \text{ mi}}{1 \text{ h}} \times \dfrac{1 \text{ h}}{\boxed{} \text{ min}}$ 570 mi/h = _____ mi/min

2. $\dfrac{108 \text{ mm}}{1 \text{ min}} \times \dfrac{\boxed{} \text{ min}}{\boxed{} \text{ s}}$ 108 mm/min = _____ mm/s

Measurement Conversions

Use the word bank to complete the steps to convert between measurement systems.

Step 1   Set up _____ for measurements you are converting.

Step 2   Write a _____ for the units you are converting.

Step 3 _____ out common units and simplify.

Word Bank

conversion factor
divide
ratio

Summary

Write 2–3 sentences to summarize the lesson.

_____

_____

_____

_____

_____

_____

_____

Rate Yourself!

Are you ready to move on? Shade the section that applies.

I have a few questions.

I'm ready to move on.

I have a lot of questions.

For more help, go online to access a Personal Tutor.

Tutor

Proportional and Nonproportional Relationships

Getting Started

Write the math and real-world definitions of constant.

• math definition _____

_____

• real-world definition _____

_____

Quick Review

Describe how you would graph point (4, 5) on a coordinate grid.

_____

_____

_____

Real-World Link

Dances The "tickets" for a school dance will be glow-in-the-dark wristbands. Students on the advisory council found the cost of the wristbands from two companies, as shown in the table.

Number of Wristbands	Total Cost ($)	
	Glow Time	Super Glow
40	14	16
60	21	18
80	28	20

1. For Glow Time wristbands, fill in the boxes to write the ratios of cost to number of wristbands in simplest form.

   $$\frac{14}{40} = \frac{\boxed{}}{20} \qquad \frac{21}{60} = \frac{\boxed{}}{\boxed{}} \qquad \frac{28}{80} = \frac{\boxed{}}{\boxed{}}$$

   Write the ratios as decimals. What do these decimals represent?

   _____

   _____

2. For Super Glow wristbands, fill in the boxes to write the ratios of cost to number of wristbands in simplest form.

   $$\frac{16}{40} = \frac{\boxed{}}{10} \qquad \frac{18}{60} = \frac{\boxed{}}{\boxed{}} \qquad \frac{20}{80} = \frac{\boxed{}}{\boxed{}}$$

   Write the ratios as decimals. _____

3. What do you notice about the simplified ratios for Glow Time and Super Glow?

   _____

   _____

Notes

Identify Proportional Relationships

Determine if the relationship shown is proportional. Justify your conclusion.

Time (h)	1	2	3	4
Distance (mi)	50	90	165	200

Use Proportional Relationships

Each table represents a proportional relationship. Complete each table. Then write the constant of proportionality in simplest form. Explain what the constant of proportionality means.

1.

Time (h)	1	2	3	4	5
Earnings ($)	12	24			

constant of proportionality: _____

2.

Tickets Ordered	2	4	6	8	10
Cost ($)	50	100			

constant of proportionality: _____

Summary

Write 2–3 sentences to summarize the lesson.

_____

_____

_____

_____

_____

_____

Rate Yourself!

Are you ready to move on? Shade the section that applies.

YES　?　NO

For more help, go online to access a Personal Tutor.

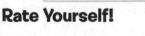 Tutor

FOLDABLES Time to update your Foldable!

Mid-Chapter Check

Vocabulary Check

1. **Be Precise** Define *ratio*. Give two examples of ratios. (Lesson 1)

_____

_____

2. Describe the difference between a proportional relationship and a nonproportional relationship. (Lesson 5)

_____

_____

_____

Skills Check and Problem Solving

Express each ratio as a fraction in simplest form. (Lesson 1)

3. 12 dolls out of 18 toys _____

4. 40 dogs to 12 cats _____

5. 35 people in 7 events _____

6. Which is less expensive per ounce, 16 ounces of Swiss cheese for $2.75 or 24 ounces of cheddar cheese for $3.25? Explain. (Lesson 2)

_____

Simplify. (Lesson 3)

7. $\dfrac{\frac{1}{2}}{\frac{2}{3}}$ _____

8. $\dfrac{\frac{1}{3}}{\frac{1}{4}}$ _____

9. $\dfrac{\frac{18}{3}}{4}$ _____

Complete each conversion. Round to the nearest hundredth. (Lesson 4)

10. 40 km/h ≈ _____ mi/min

11. 20 mi/gal ≈ _____ km/L

12. 65 mi/h ≈ _____ m/s

13. **Standardized Test Practice** Which equation can be used to represent the proportional relationship shown in the table below? (Lesson 5)

Number of Students, x	5	10	15	20	25
Total Cost ($), y	10	20	30	40	50

Ⓐ $y = \frac{1}{2}x$ Ⓑ $y = 2x$ Ⓒ $x = 2y$ Ⓓ $x = \frac{2}{x}$

21ST CENTURY CAREER
in Engineering

Start Off on the Right Foot

Use the information in the table to solve each problem.

Weight of Athlete (lb)	80	100	120	140	160
Forces Generated (lb)	200	250	300	350	400

1. For each column in the table, write the ratio of the forces generated to the weight of the athlete. Then simplify each fraction.

_____

2. Is there a constant ratio? If so, interpret its meaning.

_____

_____

3. Is there a proportional relationship between the weight of an athlete and the forces that are generated from running? Explain your reasoning.

_____

Career Project

It's time to update your career portfolio! Use the Internet or another source to research the fields of biomechanical engineering, biomedical engineering, and mechanical engineering. Write a brief summary comparing and contrasting the fields. Describe how they are all related.

_____

_____

_____

What subject in school is the most important to you? How would you use that subject in this career?

_____

_____

_____

Graphing Proportional Relationships

Getting Started

Scan Lesson 5-6 in your textbook. Predict two things you will learn about proportional relationships.

- _____
  _____
- _____
  _____

Quick Review

Write an equivalent fraction for each fraction given below.

$\frac{3}{4} =$ _____

$\frac{2}{5} =$ _____

$\frac{1}{3} =$ _____

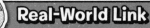

Real-World Link

Parties Mrs. Alvarez wants to order decorative cupcakes for her son's birthday celebration. The costs of dozens of cupcakes from two different bakeries are shown below.

Cupcakes Etc.	
Dozens of Cupcakes	Total Cost ($)
1	10
2	20
3	30
4	40

Mia's Cupcakes	
Dozens of Cupcakes	Total Cost ($)
1	15
2	25
3	35
4	45

1. By how much does the cost of a dozen cupcakes increase for each

   bakery? _____

2. The cost from which bakery represents a proportional relationship?

   _____

3. Graph each set of ordered pairs (dozens of cupcakes, total cost) on the coordinate plane.

4. Draw a line through each set of points. What do you

   notice about the two graphs? _____

   _____

Cupcakes

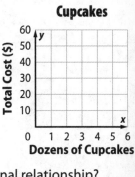

5. What can you conclude about the graph of a proportional relationship?

   _____
   _____

Notes

Identify Proportional Relationships

⊙Circle the graph(s) below that represent a proportional relationship.

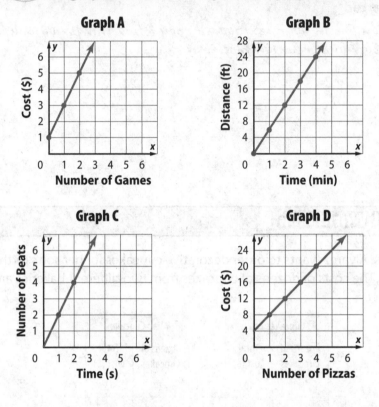

Graph A

Cost ($) vs Number of Games

Graph B

Distance (ft) vs Time (min)

Graph C

Number of Beats vs Time (s)

Graph D

Cost ($) vs Number of Pizzas

Analyze Proportional Relationships

For each proportional relationship above, what is the constant of proportionality?

Summary

Write 2–3 sentences to summarize the lesson.

Rate Yourself!

How confident are you about graphing proportional relationships? Check the box that applies.

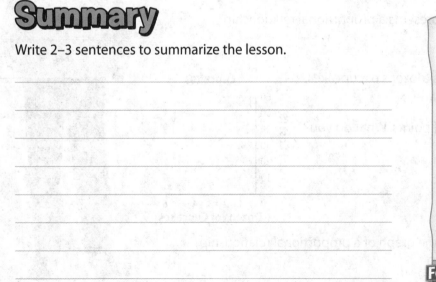

For more help, go online to access a Personal Tutor. Tutor

FOLDABLES *Time to update your Foldable!*

Solving Proportions

Getting Started

Scan Lesson 5-7 in your textbook. Write the definitions of proportion and cross products.

- proportion _____

_____

- cross products _____

Quick Review

Write the definition of equivalent ratios in your own words.

_____

_____

_____

_____

Real-World Link

Recipes You can use a recipe to make anything when you are cooking—cookies, cakes, muffins, even punch! An excellent party punch is made by mixing one gallon of fruit punch and 8 ounces of frozen sherbet.

1. Write a ratio comparing the amount of sherbet needed for one gallon of

   fruit punch. _____

2. Suppose you have three gallons of fruit punch. Complete the ratio that compares the ounces of sherbet needed to the gallons of fruit punch.

$$\frac{\boxed{}}{\boxed{}}\ \text{oz}$$
$$\text{gal}$$

3. Are the ounces of sherbet proportional to the gallons of fruit punch for one

   and three gallons? Explain. _____

_____

The recipe card shows the ingredients needed to make one dozen sugar cookies. Determine whether each ratio is proportional to the ratio of ingredients in the recipe. Justify your reasoning.

Sugar Cookies

$\frac{1}{2}$ cup butter   2 teaspoons baking powder

1 cup sugar   2 cups flour

2 eggs $\frac{3}{4}$ teaspoon vanilla

1 tablespoon milk

4. $\dfrac{3 \text{ c sugar}}{4 \text{ c flour}}$ _____

5. $\dfrac{4 \text{ T milk}}{8 \text{ t powder}}$ _____

Notes

Property of Proportions

Fill in the steps to solve $\frac{5}{16} = \frac{25}{p}$.

$\frac{5}{16} = \frac{25}{p}$ Write the proportion.

$5 \cdot p = 16 \cdot 25$ _____

$5p = 400$ _____

$\frac{5p}{5} = \frac{400}{5}$ _____

$p = 80$ _____

Use the Constant of Proportionality

Find the constant of proportionality for each situation. Then explain what it represents.

1. Mrs. Greiner bought 18 T-shirts for $144.

_____

2. Ethan earned $70 for working 8 hours.

_____

3. An airplane travels 780 miles in 4 hours.

_____

Summary

Write 2–3 sentences to summarize the lesson.

_____
_____
_____
_____
_____
_____
_____

Rate Yourself!

☐ *I understand how to solve proportions.*

▶▶ Great! You're ready to move on!

☐ *I still have questions about solving proportions.*

❚❚ No Problem! Go online to access a Personal Tutor.

Scale Drawings and Models

Getting Started

Scan Lesson 5-8 in your textbook. Write two headings you would use to make an outline of the lesson.

- _____

   _____

- _____

   _____

Quick Review

Solve for x.

$\frac{3}{10} = \frac{x}{5}$

Vocabulary Start-Up

The **scale** is the ratio of a given length on a scale drawing or model to the corresponding length on the actual object. When the scale drawing or model has the same unit of measure, the scale can be written without units. This is called the **scale factor**.

A scale of 1 inch = 2 feet can be written as 1 inch = 24 inches. So, its scale factor is 1:24.

Find each missing measure. Then find the corresponding scale factor.

Scale	Equivalent Measures	Scale Factor
1 inch = 3 feet	$\frac{1\ in.}{3\ ft} = \frac{1\ in.}{\boxed{}\ in.}$	1 : $\boxed{}$
10 centimeters = 5 meters	$\frac{10\ cm}{5\ m} = \frac{10\ cm}{\boxed{}\ cm}$	10 : $\boxed{}$ or 1 : $\boxed{}$

Real-World Link

Baseball The world's largest baseball bat is 120 feet long and 9 feet in diameter. Professional baseball players typically use bats that are 34 inches long.

Fill in the blanks to complete a proportion that could be used to find the diameter of an actual bat.

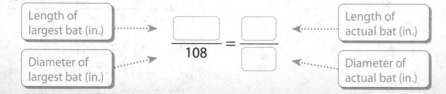

| Length of largest bat (in.) | | Length of actual bat (in.) |
| Diameter of largest bat (in.) | $\dfrac{\boxed{}}{108} = \dfrac{\boxed{}}{\boxed{}}$ | Diameter of actual bat (in.) |

Notes

Use Scale Drawings and Models

The actual measurements of a 4-room apartment are shown in the table below. Use a scale of $\frac{1}{2}$ in. = 4 ft to find length of each room on the drawing.

Room	Living Room	Kitchen	Bathroom	Bedroom
Actual Length (ft)	14	10	6	12
Drawing Length (in.)				

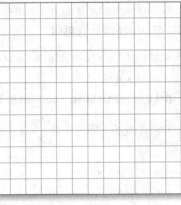

Construct Scale Drawings

Describe the process for constructing a scale drawing. Then use the information above to construct a scale drawing of one of the rooms.

_____

_____

_____

_____

_____

_____

_____

_____

Summary

Write 2–3 sentences to summarize the lesson.

_____

_____

_____

_____

_____

_____

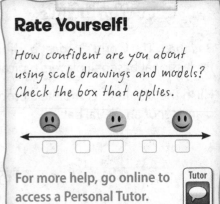

Rate Yourself!

How confident are you about using scale drawings and models? Check the box that applies.

For more help, go online to access a Personal Tutor.

Similar Figures

Getting Started

Scan Lesson 5-9 in your textbook. Write the definitions of corresponding parts and congruent.

- corresponding parts _____

_____

- congruent _____

_____

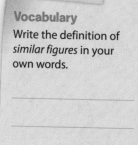

Vocabulary

Write the definition of *similar figures* in your own words.

_____

_____

_____

_____

Real-World Link

Kites The spar and spine of a kite are the sticks that hold the kite's shape. Jahmal is using the pattern at the right to make a kite with wooden dowels and nylon fabric.

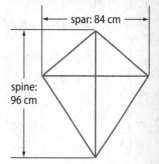

spar: 84 cm

spine: 96 cm

1. Suppose Jahmal wants to make a kite with spine and spar lengths that are two thirds the pattern's lengths. What should be the lengths of the kite's

   spine and spar? _____

2. Fill in the boxes to write a ratio that compares the spine lengths of the pattern and the kite.

   pattern (cm) ·······▸ [ ]
   ——————
   kite (cm) ·······▸ [ ]

3. Fill in the boxes to write a ratio that compares the spar lengths of the pattern and the kite.

   pattern (cm) ·······▸ [ ]
   ——————
   kite (cm) ·······▸ [ ]

4. Why are the spine lengths and spar lengths of the pattern and the kite proportional?

   _____

   _____

   _____

Notes

Similar Figures

Refer to the similar triangles below.

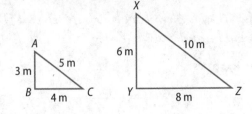

1. List the congruent angles. _____

2. List the corresponding sides. _____

Scale Factor

Explain the process you would use to find the value of *x* if *ABCD ~ EFGH*.

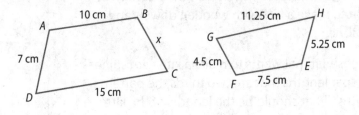

_____

_____

_____

Summary

Write 2–3 sentences to summarize the lesson.

_____

_____

_____

_____

_____

_____

_____

_____

_____

Rate Yourself!

How confident are you about similar figures? Shade the ring on the target.

I'm on target.

I need help.

For more help, go online to access a Personal Tutor.

Tutor

Indirect Measurement

Getting Started

Scan Lesson 5-10 in your textbook. List two real-world scenarios in which you would use indirect measurement.

- _____
  _____

- _____
  _____

Real-World Link

Shadows Marita is 5 feet tall and casts a shadow that is $7\frac{1}{2}$ feet long. Jacob is 6 feet tall and casts a shadow that is 9 feet long.

1. Model the situation with a labeled drawing.

2. Complete the proportion.

Marita's height ┈┈▶ ☐   ☐ ◀┈┈ Jacob's height
                      =
Marita's shadow length ┈┈▶ ☐   ☐ ◀┈┈ Jacob's shadow length

3. Find the cross products.

   _____

4. What is true about the cross products? What does this mean?

   _____
   _____

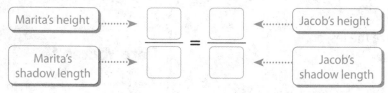

Indirect Measurement

A flagpole casts a shadow that is 32 feet long. At the same time, a shed that is 7 feet tall casts a shadow that is 17.5 feet long.

1. Model the situation with a labeled drawing.

2. Find the height of the flagpole.

Surveying Methods

The triangles below are similar. Explain how you would find the distance from Springdale to Porter. Then find the distance.

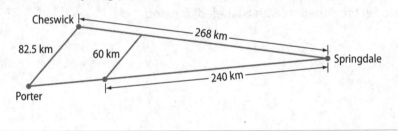

Write 2–3 sentences to summarize the lesson.

Chapter Review

Vocabulary Check

Complete the crossword puzzle using the vocabulary list at the beginning of the chapter.

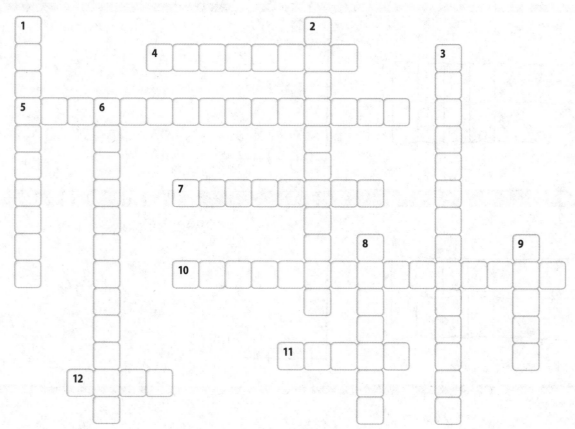

Across

4. _____ measurement allows you to use the properties of similar triangles to find missing measures

5. in the equation $y = kx$, k is the constant of _____

7. a simplified rate whose denominator is 1

10. a _____ relationship exists if two quantities do not have a constant ratio

11. the ratio of a length on a scale drawing to the corresponding length on the real object

12. a ratio of two measurements having different units

Down

1. a statement of equality of two ratios or rates

2. the ratio of the length on a scale drawing to the corresponding length with the same units on the real object

3. figures that have the same shape but not necessarily the same size

6. a _____ relationship exists if two quantities have a constant ratio or rate

8. fractions that have a fraction in the numerator, denominator, or both

9. a comparison of two quantities by division

Use Your FOLDABLES

Use your Foldable to help review the chapter.

Tape here

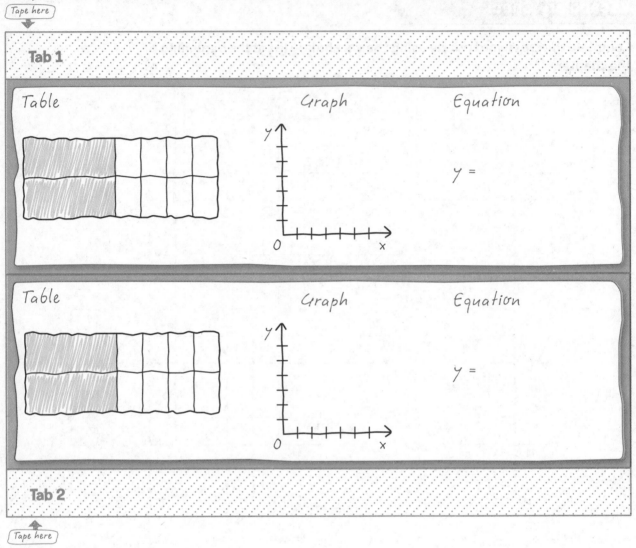

Tab 1

Table   Graph   Equation

$y =$

Table   Graph   Equation

$y =$

Tab 2

Tape here

Got it?

Determine whether each equation or graph represents a proportional or nonproportional relationship.

1. $y = 3x$

2.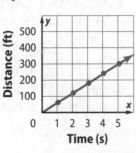

3. $y = -2x + 4$

4.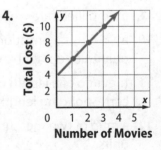

Problem Solving

1. The statistics for a school basketball league are shown in the table. Complete the table by expressing the win:loss ratio for each team as a fraction in simplest form. (Lesson 1)

Show your work.

Team	Games	Wins	Losses	Win/Loss Ratio
Bulldogs	36	30	6	
Sharks	36	26	10	
Cougars	35	23	12	
Bears	33	22	11	
Hawks	34	20	14	

2. Two stores have a certain brand of dog food on sale. Pets Etc. is selling a 17.6-pound bag for $11.99. Doggie Haven is selling a 31.8-pound bag for $21.99. Which store has the lower price per pound of dog food? Justify your reasoning. (Lesson 2) _____

_____

3. At an altitude of 250 miles, the Space Shuttle traveled at approximately 17,500 miles per hour in an orbit around Earth. In orbit, how many feet per second did the Shuttle travel? Round to the nearest foot. (Lesson 3)

_____

4. A soup recipe uses $3\frac{1}{2}$ cups of water for 8 bowls of soup. Find the amount of water Jacinda needs if she wants to make 12 bowls of soup. (Lesson 7)

_____

5. Abigail is making a scale model of her house. The front porch on Abigail's house is 25 feet long. The front porch on the model is 20 inches long. What is the scale of the model? (Lesson 8)

_____

6. Georgio wants to know how tall the tree is that grows in his backyard. His brother Mario is 4 feet tall and casts a shadow that measures 1.5 feet long. At the same time, the tree casts a shadow 4.5 feet long. How tall is the tree? (Lesson 10)

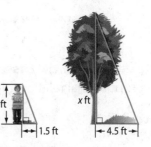

_____

Reflect

 Answering the Essential Question

Use what you learned about ratios, proportions, and similar figures to complete the graphic organizer.

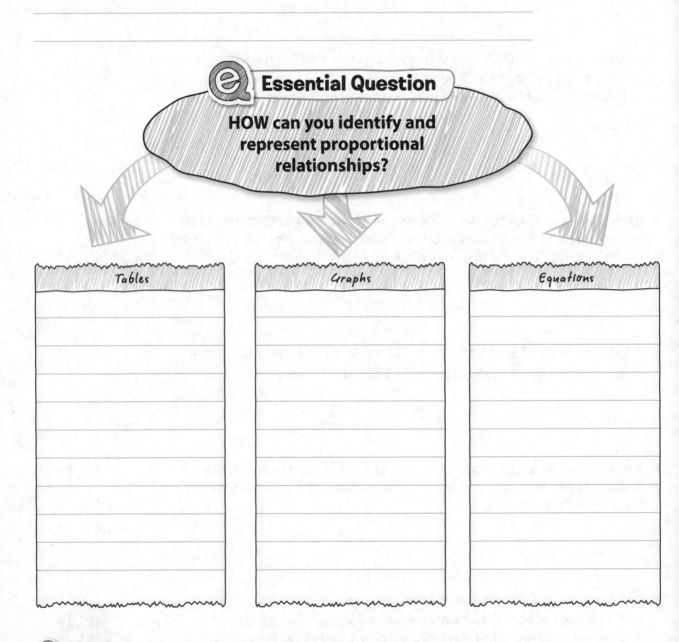

Essential Question

HOW can you identify and represent proportional relationships?

Tables

Graphs

Equations

Answer the Essential Question. HOW can you identify and represent proportional relationships?

Chapter 6
Percents

Chapter Preview

Vocabulary

compound interest	percent error	percent proportion
discount	percent of change	principal
interest	percent of decrease	selling price
markup	percent of increase	simple interest
percent equation		

Key Concept Activity

Look through the chapter. Write one or two key concepts from each lesson. Later, you can use this as a study tool for your chapter review.

Lesson	Key Concept(s)
6-1 Using the Percent Proportion	
6-2 Find Percent of a Number Mentally	
6-3 Using the Percent Equation	
6-4 Percent of Change	
6-5 Discount and Markup	
6-6 Simple and Compound Interest	

Quick Review
CCSS

Common Core Review 6.EE.7, 7.RP.2

Example 1

Solve $7.5x = 30$.

$7.5x = 30$	Write the equation.
$\dfrac{7.5x}{7.5} = \dfrac{30}{7.5}$	Division Property of Equality
$x = 4$	Simplify.

Example 2

Solve $\dfrac{3}{5} = \dfrac{n}{20}$.

$\dfrac{3}{5} = \dfrac{n}{20}$	
$3 \cdot 20 = 5n$	Cross products
$60 = 5n$	Multiply.
$\dfrac{60}{5} = \dfrac{5n}{5}$	Division Property of Equality
$12 = n$	Simplify.

Quick Check

One-Step Equations Solve each equation. Check your solution.

1. $25x = 50$ _____

2. $28 = 8a$ _____

3. $0.1n = 0.5$ _____

Show your work.

Proportions Solve each proportion.

4. $\dfrac{2}{15} = \dfrac{n}{45}$ _____

5. $\dfrac{3}{2} = \dfrac{n}{10}$ _____

6. $\dfrac{11}{50} = \dfrac{n}{200}$ _____

7. The wait time to ride the Ferris wheel at a state fair is 15 minutes when 75 people are in line. At this rate, what is the wait time when 200 people are in line? _____

How Did You Do?

Which problems did you answer correctly in the Quick Check?
Shade those exercise numbers below.

① ② ③ ④ ⑤ ⑥ ⑦

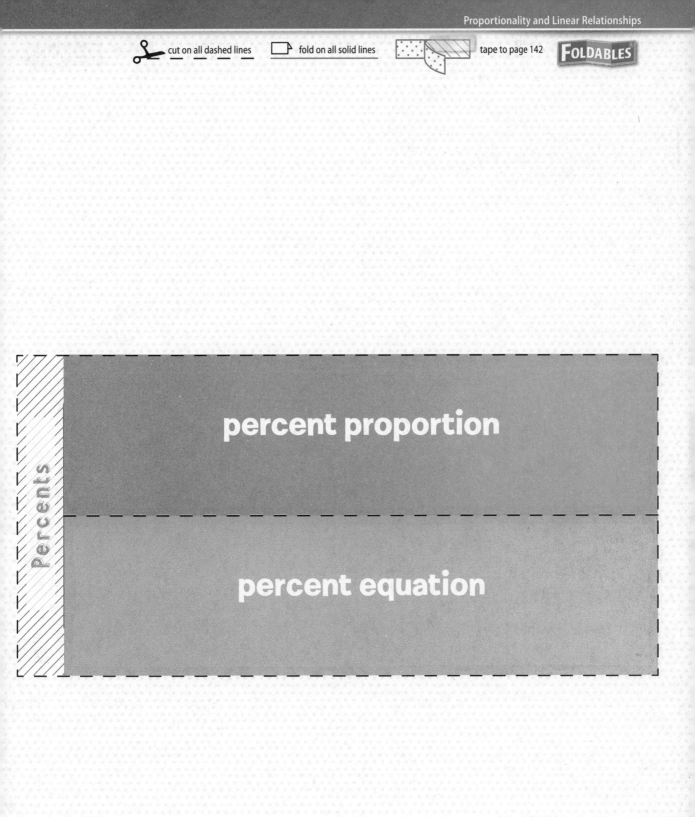

Percents

percent proportion

percent equation

1 Cut out the Foldable above.

2 Place your Foldable on page 142.

3 Use the Foldable throughout this chapter to help you learn about percents.

Definition

Definition

page 142

Using the Percent Proportion

Getting Started

Scan Lesson 6-1 in your textbook. Predict two things you will learn about the percent proportion.

- _____
- _____

Vocabulary

Write the definition of *proportion* in your own words.

_____

_____

_____

Real-World Link

Snacks With four different kinds of fruit, this healthy fruit salad recipe is the perfect lunch box or after school snack!

> **Fruit Salad**
> 2 cups pineapple
> 1 cup blueberries
> 3 cups grapes
> 2 cups strawberries

1. What is the total amount of ingredients needed to make one batch of fruit salad? _____

2. Write the ratio comparing the cups of grapes to the total cups needed.

$$\frac{part}{whole} = \frac{\boxed{}c}{\boxed{}c}$$

3. Write the fraction from Exercise 2 as a decimal. _____

4. Solve the proportion $\frac{3}{8} = \frac{p}{100}$. _____

5. How does your answer for Exercise 4 compare to your answer for Exercise 3? _____

_____

6. What does the ratio $\frac{p}{100}$ represent? _____

_____

7. What percent of the trail mix are the sunflower seeds? _____

Notes

Percent Proportion

Label the diagram below with the terms *part, whole*, and *percent*.

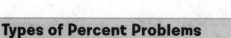

$$\frac{3}{8} = \frac{n}{100}$$

Types of Percent Problems

Complete the table to show an example of each type of percent problem. The first one is done for you.

Type	Example	Proportion
Find the Percent	6 is what percent of 12?	$\frac{6}{12} = \frac{p}{100}$
Find the Part		
Find the Whole		

Use the percent proportion to find each value.

1. 36 is what percent of 80?

2. What number is 15% of 220?

Summary

Write 2–3 sentences to summarize the lesson.

Rate Yourself!

How confident are you about using the percent proportion? Shade the ring on the target.

For more help, go online to access a Personal Tutor.

FOLDABLES Time to update your Foldable!

Find Percent of a Number Mentally

Getting Started

Scan Lesson 6-2 in your textbook. List two headings you would use to make an outline of the lesson.

- _____
  _____

- _____
  _____

Quick Review

Write each percent as a decimal and as a fraction in simplest form.

75% _____

60% _____

10% _____

Real-World Link

Thrill Rides Do you enjoy thrill rides? *Power Tower* is a thrill ride that is 300 feet tall. Two of the towers blast riders upward and two towers drop riders downward. In both cases, passengers travel 80% of the ride's total height.

1. How would you find how far the riders travel on the ride?

   _____

2. *Compatible numbers* are numbers that are easy to multiply or divide mentally. Explain how you could use compatible numbers to mentally find 80% of 300.

   _____

   _____

3. Write 80% as a decimal and as a fraction in simplest form.

   Decimal   Fraction

   $$80\% = \boxed{} = \frac{\boxed{}}{\boxed{}}$$

4. Is it easier to use the decimal form of 80% or the fractional form of 80% to find 80% of 300? Explain. _____

   _____

   How far are the riders blasted upward on the ride? _____

5. Describe another method you could use to mentally find 80% of 300. _____

   _____

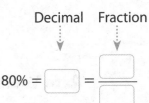

Notes

Find Percent of a Number Mentally

Describe a method you could use to mentally find 40% of 700. Then use that method to find the answer.

_____

_____

Complete the table to show some common percent-fraction equivalents.

Percent-Fraction Equivalents		
25% = ___	20% = ___	10% = ___
50% = ___	40% = ___	30% = ___
75% = ___	60% = ___	70% = ___
100% = ___	80% = ___	90% = ___

Estimate with Percents

Describe a strategy to estimate each value. Use a different strategy each time.

1. 150% of 98 _____

_____

2. 76% of 160 _____

_____

3. $\frac{1}{2}$% of 280 _____

Summary

Write 2–3 sentences to summarize the lesson.

_____

_____

_____

_____

_____

_____

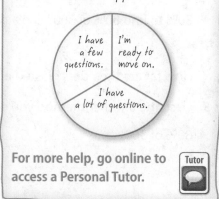

Rate Yourself!

Are you ready to move on? Shade the section that applies.

I have a few questions.

I'm ready to move on.

I have a lot of questions.

For more help, go online to access a Personal Tutor.

Tutor

Using the Percent Equation

Getting Started

Scan Lesson 6-3 in your textbook. List two real-world scenarios in which you would use percent.

- _____
  _____

- _____
  _____

Vocabulary Start-Up

You have used the percent proportion to solve problems involving percent. You can also use a **percent equation**.

Label the diagram with the terms *part*, *whole*, and *percent* to derive the percent equation from the percent proportion.

$$\frac{\text{part}}{\text{whole}} = \frac{p}{100}$$

$$\frac{\boxed{}}{\boxed{}} = \boxed{}$$

$$\frac{\boxed{}}{\boxed{}} \cdot \text{whole} = \boxed{} \cdot \text{whole}$$

$$\boxed{} = \boxed{} \cdot \boxed{}$$

Real-World Link

Elephants An adult male African Elephant weighs 15,000 pounds. An adult female African Elephant weighs 53% of the male elephant's weight. How would you write an equation to find the weight of the female elephant?

$$\boxed{} = 0.53 \cdot \boxed{}$$

The maximum weights for male and female adult Asian Elephants are 11,000 and 6000 pounds, respectively. How would you write an equation to determine what percent of 11,000 is 6000?

$$\boxed{} = \boxed{} \cdot 6000$$

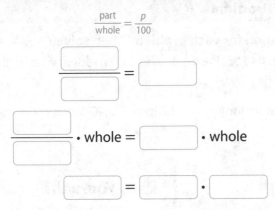

Notes

Percent Equation

Complete the graphic organizer. Write the type of percent problem using the terms *part, whole,* and *percent*. Then solve using the percent equation. The first one is done for you.

> Percent Equation
> part = percent x whole

82.5 is what percent of 500?

Find the percent.

$82.5 = p \cdot 550$

$\dfrac{82.5}{550} = \dfrac{550p}{550}$

$0.15 = p$

$p = 15\%$

What is 15% of 142?

Find the _____.

_____

_____

_____

14 is 20% of what number?

Find the _____.

_____

_____

_____

_____

Solve Real-World Problems

Write a percent equation for each problem. Then solve.

1. A boxed DVD set costs \$36. If a $6\frac{1}{2}\%$ sales tax is added, what is the total cost? _____

2. The cost of a haircut including a 20% tip is \$50.40. What is the cost of the haircut alone? _____

Summary

Write 2–3 sentences to summarize the lesson.

_____

_____

_____

_____

_____

_____

Rate Yourself!

How confident are you about use the percent equation? Shade the ring on the target.

I'm on target.

I need help.

For more help, go online to access a Personal Tutor.

Tutor

FOLDABLES *Time to update your Foldable!*

Mid-Chapter Check

Vocabulary Check

1. Be Precise Describe the parts of the percent proportion. (Lesson 1)

_____

_____

_____

Fill in the blank.

2. The percent equation states that the part equals the percent times

the _____. (Lesson 3)

Skills Check and Problem Solving

Use the percent proportion or percent equation to solve each problem. (Lessons 1 and 3)

3. 36 is what percent of 50? _____

4. What percent of 7 is 21? _____

5. What is 83% of 16? _____

6. 30 is what percent of 75? _____

7. The Dragons Soccer Team is having a pizza party and voted on the toppings the players preferred. Of the 25 players, what

percent prefer cheese? (Lesson 1) _____

Toppings	Number of Players
Cheese	9
Pepperoni	8
Sausage	3
Hawaiian	5

Find the percent of each number mentally. (Lesson 2)

8. 1% of 80 _____

9. 25% of 160 _____

10. $87\frac{1}{2}$ % of 56 _____

11. Standardized Test Practice A beanbag chair costs $84. If a 7.25% sales tax is added, what is the total cost? (Lesson 3)

Ⓐ $6.09

Ⓒ $89.88

Ⓑ $60.90

Ⓓ $90.09

21ST CENTURY CAREER
in Video Game Design

All Fun and Games

Use the information in the circle graph and the table to solve the problems below.

1. How many of the top 20 video games sold were sports games? _____

2. Out of the top 20 video games sold, how many more music games were there than racer games? _____

3. In Week 1, the total sales for a video game were $2,374,136. What percent of the total sales was from the United States? Round to the nearest whole percent. _____

4. In Week 3, what percent of the total sales in the United States and Japan were Japan's sales? Round to the nearest whole percent. _____

5. Did the sales in the United States have a greater percent of the two countries' total video game sales in Week 1 or in Week 2? Explain.

_____

_____

_____

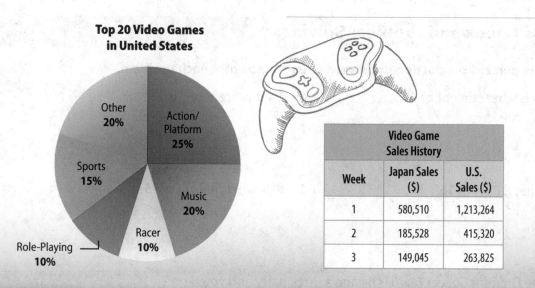

Top 20 Video Games in United States

Other 20%
Action/Platform 25%
Sports 15%
Music 20%
Role-Playing 10%
Racer 10%

Video Game Sales History		
Week	Japan Sales ($)	U.S. Sales ($)
1	580,510	1,213,264
2	185,528	415,320
3	149,045	263,825

Career Project

It's time to update your career portfolio! Choose one of your favorite video games. Make a list of what you think are the best features of the game. Then describe any changes that you, as a video game designer, would make to the game.

_____

_____

_____

_____

_____

List the strengths you have that would help you succeed in this career.

- _____
- _____
- _____
- _____
- _____

Percent of Change

Getting Started

Scan Lesson 6-4 in your textbook. Write the definitions of percent of change and discount.

- percent of change _____

_____

- discount _____

_____

Quick Review

Use the percent equation to find each value.
3 is what percent of 10?

_____

18 is 25% of what

number? _____

40% of 50 is what

number? _____

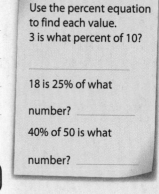

Real-World Link

Movies Movies sure have come a long way! The first known motion picture was filmed in 1888 and lasted for only 2.11 seconds. Today, we watch motion pictures that last an average of about two hours. The table below shows the average cost of a ticket in several different years.

Average Cost of a Movie Ticket	
Year	Cost ($)
1992	4.15
2002	5.80
2012	7.93

1. How much more did a ticket cost in 2012 than in 2002? _____

2. Write the ratio $\dfrac{\text{cost increase from 2002 to 2012}}{\text{cost in 2002}}$. Then write the ratio as a percent. Round to the nearest tenth.

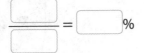

3. Write the ratio $\dfrac{\text{cost increase from 1992 to 2012}}{\text{cost in 1992}}$. Then write the ratio as a percent. Round to the nearest whole percent.

4. Suppose the percent in Exercise 3 was 100%. Explain the meaning of 100% in this situation. _____

_____

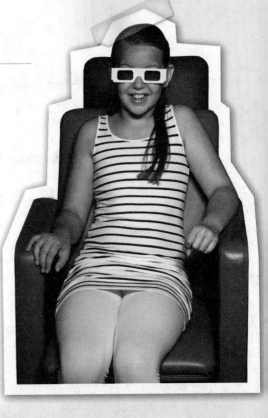

Notes

Percent of Change

Fill in the blanks to find the percent of change. Round to the nearest whole percent.

Last fall, 30 students were in Juanita's after-school club. This fall, there are 58 students.

$$\dfrac{\text{amount of change}}{\text{original amount}} = \dfrac{\boxed{}}{\boxed{}} \qquad \text{Substitute.}$$

$$\approx \boxed{} \qquad \text{Write as a decimal.}$$

$$\approx \boxed{}\% \qquad \text{Write as a percent.}$$

There was a _____ % _____ in students.

Percent Error

Fill in the blanks to find the percent error.

Mrs. Harris estimates that she will spend $65 on school supplies for her children. She actually spends $90.

amount of error: $\boxed{}$

actual amount: $\boxed{}$

percent error: $\dfrac{\boxed{}}{\boxed{}} \approx \boxed{}$ or $\boxed{}$ %

Summary

Write 2–3 sentences to summarize the lesson.

_____

_____

_____

_____

_____

_____

Rate Yourself!

$\boxed{}$ *I understand how to find the percent of change.*

▶▶ Great! You're ready to move on!

$\boxed{}$ *I still have questions about percent of change.*

⏸ No Problem! Go online to access a Personal Tutor.

Discount and Markup

Getting Started

Scan Lesson 6-5 in your textbook. List two headings you would use to make an outline of the lesson.

- _____
  _____

- _____
  _____

Vocabulary

Check the boxes of words you already know.

☐ markup

☐ selling price

☐ discount

Real-World Link

School Supplies Imani went shopping for school supplies during a sales tax holiday. She purchased 1 binder, 4 folders, 3 notebooks, and 2 packs of pens. The prices are shown below.

Item	Cost ($)
3-ring binder	2.99
folder	0.79
notebook	1.49
pack of pencils	0.99
pack of pens	1.29

1. Write a numerical expression to find the total cost of the school supplies purchased by Imani.

   _____

2. Use the expression you wrote in Exercise 1 to find the total cost of the

   school supplies. _____

3. Suppose there is normally a 7% sales tax added to the total cost of the items. Find the amount of tax Imani would have had to pay without the

   sales tax holiday. Round to the nearest cent. _____

4. Write a numerical expression to find how much Imani would have spent on school supplies without the sales tax holiday. Then find the total amount.

   Expression: _____

   Total Amount with Tax: _____

Notes

Using Markup

A store pays \$75 for an eReader and its markup is 50%. Explain how you would find the selling price using both methods. Then find the selling price of the eReader.

Find Markup Amount First

Find Total Percent First

Selling Price

Using Discount

1. How is discount similar to markup? How is it different?

2. A pair of athletic shoes has an original price of \$145. The shoes are on sale for 20% off the original price. Find the sale price.

3. A bracelet sells for \$36. It is on sale for 25% off the original price. What will the customer pay to the nearest cent for the bracelet if there is a 6.5% sales tax?

Summary

Write 2–3 sentences to summarize the lesson.

Rate Yourself!

Are you ready to move on? Shade the section that applies.

YES ? NO

For more help, go online to access a Personal Tutor.

Simple and Compound Interest

Getting Started

Scan Lesson 6-6 in your textbook. Predict two things you will learn about finding interest earned or owed.

* _____

_____

* _____

_____

Quick Review

Write each percent as a decimal.

$6\frac{1}{2}$% _____

2.75% _____

$1\frac{1}{10}$% _____

Vocabulary Start-Up

Interest is the amount of money paid or earned for the use of money by a bank or other financial institution. The **principal** is the amount of money invested or borrowed. Label the variables in the simple interest formula shown below with the terms *interest*, *principal*, *rate*, and *time*.

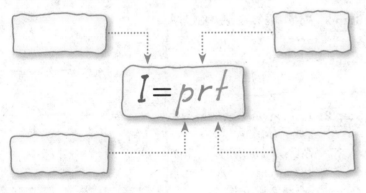

$$I = prt$$

Real-World Link

Cars Typically when you buy a car, you pay a certain amount upfront and get a loan to cover the rest. The Ramirez family borrowed $6000 at 3.25% for 4 years to buy a new car. Use the formula above to find the amount of simple interest the family will pay on the loan.

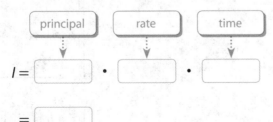

$I =$ [principal] · [rate] · [time]

$=$ [ ]

Notes

Simple Interest

Circle the correct equation to find the interest earned for each situation.

1. $1500 at 3.5% for 2 years $I = 1500 \cdot 3.5 \cdot 2$ $I = 1500 \cdot 0.035 \cdot 2$

2. $3000 at 2% for 12 months $I = 3000 \cdot 0.02 \cdot 12$ $I = 3000 \cdot 0.02 \cdot 1$

3. When finding simple interest, what are two things you need to remember?

_____

_____

Compound Interest

Complete the diagram to find the amount of interest earned when $800 is invested at 6.75% compounded annually for 2 years.

Step 1 Find the interest earned after 1 year. ┈┈┈▶ [ ]

Step 2 Add the interest earned after the first year to the principal. Then find the interest earned after 2 years. ┈┈┈▶ [ ]

Summary

Write 2–3 sentences to summarize the lesson.

_____

_____

_____

_____

_____

_____

_____

Rate Yourself!

How confident are you about finding simple and compound interest? Check the box that applies.

☹   ☹   ☺

☐ ☐ ☐ ☐ ☐

For more help, go online to access a Personal Tutor.

Tutor

Chapter Review

Vocabulary Check

Unscramble each of the clue words. After unscrambling all of the terms, use the numbered letters to find another vocabulary term from the chapter.

LESGLNI RIEPC

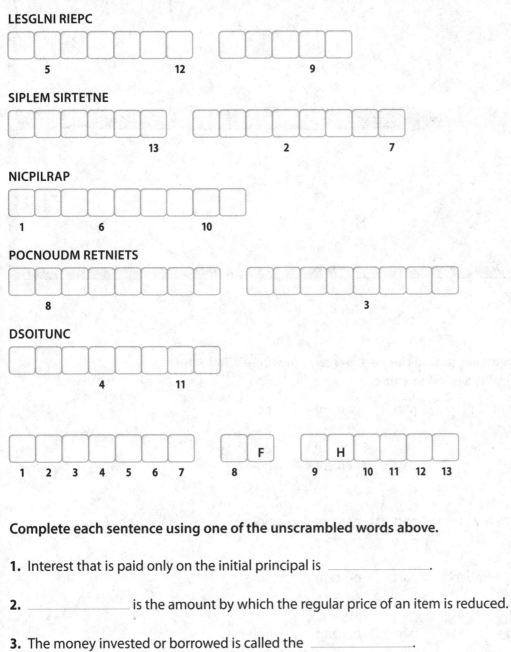

□□□□□□□ □□□□□
  5   12   9

SIPLEM SIRTETNE

□□□□□□ □□□□□□□□
       13   2   7

NICPILRAP

□□□□□□□□□
1   6   10

POCNOUDM RETNIETS

□□□□□□□□ □□□□□□□□
  8   3

DSOITUNC

□□□□□□□□
    4   11

□□□□□□□ □[F]□ □[H]□□□□
1 2 3 4 5 6 7   8   9   10 11 12 13

Complete each sentence using one of the unscrambled words above.

1. Interest that is paid only on the initial principal is _____.

2. _____ is the amount by which the regular price of an item is reduced.

3. The money invested or borrowed is called the _____.

4. _____ is interest that is paid on the initial principal and on interest earned in the past.

5. The _____ is the amount of money a customer pays for an item.

Use Your FOLDABLES

Use your Foldable to help review the chapter.

Tape here

Percents

Examples

Examples

Got it?

Determine whether you are to find the *part, percent,* **or** *whole.* **Circle the correct term. Then find the missing value.**

1. 70 is what percent of 280?   part   percent   whole

_____

2. What is 118% of 19?   part   percent   whole

_____

3. What percent of 800 is 2?   part   percent   whole

_____

4. 126 is 30% of what number?   part   percent   whole

_____

5. 25% of what number is 10?   part   percent   whole

_____

6. What number is 4% of 30?   part   percent   whole

_____

Problem Solving

1. The table shows the eye colors of the students in Mr. Lehman's class. Use the percent proportion to find the percent of students with blue eyes.

   (Lesson 1) _____

Eye Color	Number of Students
Blue	9
Brown	5
Green	6
Hazel	4

   show your work.

2. The bill for the Morgan family at a restaurant was $62.14. Mr. Morgan would like to leave their server a 20% tip. About how much of a tip will the server

   receive? (Lesson 2) _____

3. In a community of 3200 homes, 72% of the households participate in the community recycling program. How many households is this?

   (Lesson 3) _____

4. The figures for home sales for 2013 and 2014 in Shore County are shown. (Lesson 4)

   a. Find the percent of change in the homes sold. Round to the nearest tenth. _____

   b. Find the percent of change in the average selling price. Round to the nearest tenth. _____

Shore County Home Sales		
Year	Number of Homes Sold	Average Selling Price
2013	18,328	$220,988
2014	19,226	$219,748

5. Inez is shopping for clothes at a department store. Before tax, her bill is $95. (Lesson 5)

   a. She has a coupon to receive an additional 25% off her total purchase. What is the total cost of her items before tax? _____

   b. After she receives the discount, how much will her total bill be if there is a 7.95% sales tax? _____

6. Marcus borrowed $4000 to buy a car. If his monthly payments are $184.17 for 2 years, what is the simple interest rate for the loan? (Lesson 6)

   _____

Reflect

Answering the Essential Question

Use what you learned about percent to complete the graphic organizer. For each situation, circle an arrow to show if the final amount would be greater or less than the original amount. Then write a real-world percent problem and an equation that models it.

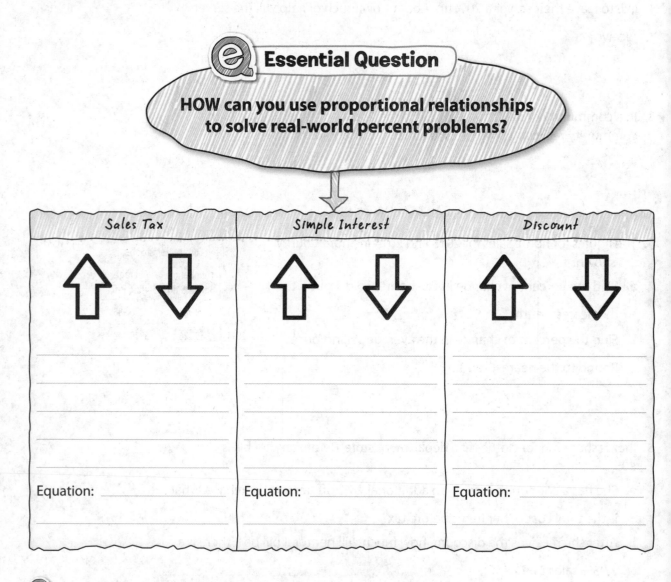

Essential Question

HOW can you use proportional relationships to solve real-world percent problems?

Sales Tax	Simple Interest	Discount
⬆ ⬇	⬆ ⬇	⬆ ⬇
Equation:	Equation:	Equation:

Answer the Essential Question. HOW can you use proportional relationships to solve real-world percent problems?

_____

_____

Chapter 7
Algebraic Expressions

 **Vocabulary**

coefficient	factor	simplest form
constant	factored form	simplifying the expression
Distributive Property	like terms	term
equivalent expressions	linear expression	

Vocabulary Activity

Use the glossary to find the definitions of the terms below. Then draw a line to match each term with the correct definition.

1. equivalent expressions

2. term

3. coefficient

4. like terms

5. constant

6. simplest form

7. simplify the expression

8. linear expression

a. The numerical part of a term that contains a variable.

b. An algebraic expression that has no like terms and no parentheses.

c. Expressions that contain the same variables to the same power.

d. Expressions that have the same value.

e. A term without a variable.

f. An algebraic expression in which the variable is raised to the first power.

g. To use distribution to combine like terms.

h. The different parts of an algebraic expression that are separated by addition or subtraction signs.

Try the Quick Check below.
Or, take the Online Readiness Quiz.

Check ✓

CCSS **Quick Review**   **Common Core Review** 7.NS.1, 7.NS.2

Example 1

Find 7(−2).

7(−2) = −14   The factors have different signs. The product is negative.

Example 2

Write 8 − 12 as an addition expression. Then find the value of the expression.

8 − 12 = 8 + (−12)   To subtract 12, add −12.

= −4   Simplify.

Quick Check

Multiplying Integers Find each product.

1. 3(−3) = _____

2. −4(2) = _____

3. −7(−4) = _____

4. −4 · 5 = _____

5. −11(−8) = _____

6. 9(−6) = _____

7. The price of a stock decreased $2 each day for 5 consecutive days. Write a multiplication expression for the total change in the value of the stock over the five-day period. Then find the total change. _____

Show your work.

Subtracting Integers Write each subtraction expression as an addition expression. Then find the value of the expression.

8. 4 − 10 _____

9. −11 − 5 _____

10. Student Council spent $178 on decorations and $110 on snacks for a dance. Write an addition expression for the amount remaining in the budget if Student Council initially had $593. Then find the amount remaining.

_____

How Did You Do?

Which problems did you answer correctly in the Quick Check? Shade those exercise numbers below.

1 2 3 4 5 6 7 8 9 10

cut on all dashed lines   fold on all solid lines   tape to page 162   **FOLDABLES**

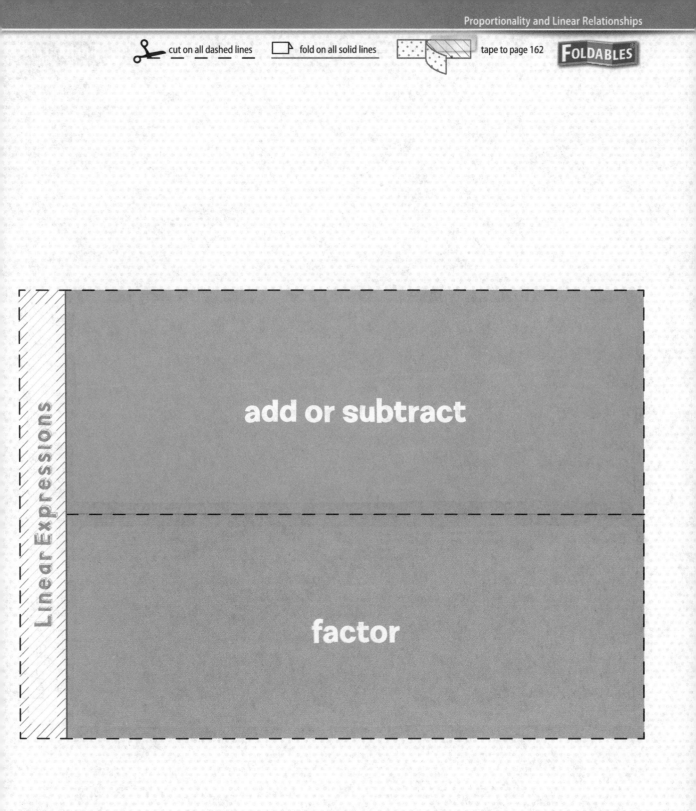

Linear Expressions

add or subtract

factor

FOLDABLES
Study Organizer

1 Cut out the Foldable above.

2 Place your Foldable on page 162.

3 Use the Foldable throughout this chapter to help you learn about linear expressions.

Examples

Examples

page 162

The Distributive Property

Getting Started

Scan Lesson 7-1 in your textbook. List two headings you would use to make an outline of the lesson.

- _____
  _____

- _____
  _____

Real-World Link

Entertainment The table shows the cost of different activities at the Newport Aquarium. Jairo and his three cousins each paid for admission and went on the Behind the Scenes tour.

Activity	Cost per Person ($)
Admission	25
Penguin Encounters	23
Behind the Scenes Tour	15

1. Complete the expression that represents the cost of four admissions and four Behind the Scenes tours.

   ☐ · 25 + ☐ · 15

2. Complete the expression that represents four times the sum of the cost of one admission and one Behind the Scenes tour.

   ☐ (☐ + ☐)

3. Evaluate the expressions in Exercises 1 and 2. What do you notice?

   _____

4. Suppose Jairo and his three cousins also went to Penguin Encounters. Circle the expression that represents the total cost of their visit to the Newport Aquarium.

   4 · 25 · 15 · 23   4 · 25 + 4 · 15 + 4 · 23

5. Write another expression that has the same value as the expression you circled in Exercise 4.

Notes

Numerical Expressions

Complete each expression using the Distributive Property.

1. $5(3 + 4) = 5 \cdot 3 + 5 \cdot \boxed{}$

2. $6(7 - 1) = 6 \cdot \boxed{} - 6 \cdot 1$

3. $\frac{1}{2}(8 - 3) = \frac{1}{2} \cdot \boxed{} - \frac{1}{2} \cdot \boxed{}$

4. $(6 + 3)0.7 = 6 \cdot \boxed{} + 3 \cdot \boxed{}$

5. Complete the sentence below to describe how to use the Distributive Property.

   When using the Distributive Property, _____

   _____

Algebraic Expressions

Use a model to rewrite $2(x + 3)$ as an equivalent expression.

Summary

Write 2–3 sentences to summarize the lesson.

_____

_____

_____

_____

_____

_____

_____

Rate Yourself!

Are you ready to move on? Shade the section that applies.

YES   ?   NO

For more help, go online to access a Personal Tutor.

Tutor

Simplifying Algebraic Expressions

Getting Started

Scan Lesson 7-2 in your textbook. Predict two things you will learn about algebraic expressions.

- _____
  _____

- _____
  _____

Vocabulary

Check the boxes of the vocabulary terms that you already know.

☐ algebraic expression

☐ coefficient

☐ constant

☐ like terms

☐ term

Vocabulary Start-Up

When addition or subtraction signs separate an algebraic expression into parts, each part is a **term**. In this chapter, we will work only with terms with an exponent of 1. In this case, **like terms** are terms that contain the same variables, such as 2*n* and 5*n* or 6*xy* and 4*xy*.

Underline the variable(s) in each term. Then cross out the term in each group that is not like the other two terms.

5x	**4y**	**7x**
3c	**−2c**	**6d**
2mn	**3np**	**6np**
5rs	**−5rs**	**5st**

Real-World Link

Recycling Suppose one class collected *p* pounds of recyclables and a second class collected six more pounds of recyclables more than the first class.

1. Write an expression to represent the pounds of recyclables collected by each class.

   first class: _____   second class: _____

2. Write an addition expression to represent the total pounds of recyclables collected by both classes.

   _____

Notes

Parts of Algebraic Expressions

1. Identify the parts of the algebraic expression below. Write *coefficient*, *constant*, or *term*.

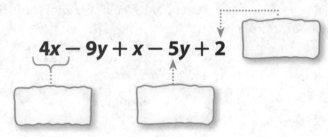

$$4x - 9y + x - 5y + 2$$

2. Identify two pairs of like terms. _____

3. What is the coefficient of the third term? _____

Simplify Algebraic Expressions

Fill in the boxes to write each expression in simplest form.

4. $4x + 3x = \boxed{}\, x$

5. $10 + 4.5y + 6.5y = 10 + \boxed{}\, y$

6. $15a + 6b - a + 2b = \boxed{}\, a + \boxed{}\, b$

7. $3t + 1 + 8t - 6 = \boxed{}\, t - \boxed{}$

8. $2m - 4k + 3 - 8m + 2 = \boxed{}\, m - \boxed{}\, k + \boxed{}$

Summary

Write 2–3 sentences to summarize the lesson.

_____

_____

_____

_____

_____

_____

_____

Rate Yourself!

How confident are you about simplifying algebraic expressions? Shade the ring on the target.

I'm on target.

I need help.

For more help, go online to access a Personal Tutor.

Tutor

Adding Linear Expressions

Getting Started

Scan Lesson 7-3 in your textbook. List two real-world scenarios in which you would add linear expressions.

- _____
  _____

- _____
  _____

Real-World Link

Engineering A *trebuchet* is a medieval catapult that was used to hurl large projectiles at castle walls. Rashawn and Jordan are building their own trebuchets to see which one can launch objects farther. Each student is using the amounts of wood shown in the table.

Length of Wood (in.)	Number of Pieces
10	4
12	3
16	1

1. Complete the expressions below to represent the pieces of wood that each student needs.

   Rashawn:

   ☐ 16-inch piece + ☐ 12-inch pieces + ☐ 10-inch pieces

   Jordan:

   ☐ 16-inch piece + ☐ 12-inch pieces + ☐ 10-inch pieces

2. Complete the expression below to represent the total number of pieces of wood that the students need.

   ☐ 16-inch pieces + ☐ 12-inch pieces + ☐ 10-inch pieces

3. Explain how you found the solution for Exercise 2.

   _____

   _____

4. Rashawn uses 2 rubber balls and Jordan uses 5 rubber balls to test their trebuchets. Then they each use *x* clay balls to test their trebuchets. The model below represent the total number of balls that they use. Write an expression in simplest form to represent the model.

_____

Notes

Add Linear Expressions

1. Draw a model to represent $(2x + 1) + (3x + 2)$.

2. Use the model in Exercise 1 to add $(2x + 1) + (3x + 2)$. _____

3. How is adding linear expressions with algebra tiles similar to adding linear expressions by combining like terms?

_____

_____

Find Perimeter

Let s represent the length of the first side of a triangle. The length of the second side of the triangle is three more than the first side. The length of the third side is four less than 1.5 times the first side.

4. Write an expression for the length of the second and third sides of the triangle.

second side: _____

third side: _____

5. Write and simplify an expression to find the perimeter of the triangle.

_____

Summary

Write 2–3 sentences to summarize the lesson.

_____

_____

_____

_____

_____

_____

Rate Yourself!

How well do you understand adding linear expressions? Circle the image that applies.

Clear   Somewhat   Not So
              Clear   Clear

For more help, go online to access a Personal Tutor. [Tutor]

 Time to update your Foldable!

Mid-Chapter Check

Vocabulary Check

1. **Be Precise** Define *equivalent expressions*. Give an example of two equivalent expressions. (Lesson 1)

_____

_____

2. Fill in the blank in the sentence below with the correct term. (Lesson 2)

A _____ is the numerical factor of a multiplication expression like 4*x*.

Skills Check and Problem Solving

Simplify each expression. (Lessons 1 and 2)

3. $8(x + 3) =$ _____

4. $4(x - 5) =$ _____

5. $9y + 3 - y =$ _____

6. $6.5(m + 2) - 2m =$ _____

Add. Use models if needed. (Lesson 3)

7. $(2x + 6) + (5x + 4) =$ _____

8. $(-6x + 3) + (-2x + 7) =$ _____

9. $(4x - 5) + (8x - 6) =$ _____

10. $3(x + 4) + (2x + 6) =$ _____

11. Write an expression in simplest form that represents the perimeter of the rectangle shown. Then find the perimeter if $x = 5$. (Lesson 3) _____

2x − 5 in.

6 in.   6 in.

2x − 5 in.

12. **Standardized Test Practice** Lucita works at a bookstore and earns $9.50 per hour. She works 3 hours on Friday and 7 hours on Saturday. Which expression does *not* represent her wages for those days? (Lesson 1)

Ⓐ 9.5(3 + 7)

Ⓒ 9.5(3) + 9.5(7)

Ⓑ 10(9.5)

Ⓓ 9.5(3)(7)

A Thrilling Ride

The note cards show statistics for three popular roller coasters at one amusement park. Use the information on the note cards to solve each problem.

1. London rode each roller coaster twice. Write two expressions that could be used to find the amount of time London spent riding the coasters. Then find the total amount of time.

   _____

2. Each time London rode the Millennium Force, she waited in line for the same amount of time *t*. Write an expression to show the

   total time she spent in line and on the ride. _____

3. Write and simplify an expression to show the total length of the Magnum XL-200 and the Top Thrill Dragster.

   _____

   If the Top Thrill Dragster is 2800 feet long, how long is the Magnum

   XL-200? _____

4. Write and simplify an expression to represent the average top speed of the three roller coasters.

   _____

   _____

Millennium Force
- Top Speed: $(s + 21)$ mph
- Max Height: 310 ft
- Length: $(2x + 995)$ ft
- Duration: 2.3 min

Top Thrill Dragster
- Top Speed: $(s + 48)$ mph
- Max Height: 420 ft
- Length: x ft
- Duration: 0.5 min

Magnum XL-200
- Top Speed: s mph
- Max Height: 205 ft
- Length: $(2x - 494)$ ft
- Duration: 2 min

Career Project

It's time to update your career portfolio! Describe a roller coaster that you, as a roller coaster designer, would create. Include the height and angle of the tallest drop, the total length, maximum speed, number of loops and tunnels, and color scheme. Be sure to include the name of your roller coaster.

_____

_____

_____

_____

What problem-solving skills might you use as a roller coaster designer?
- _____
- _____
- _____
- _____
- _____

Subtracting Linear Expressions

Getting Started

Scan Lesson 7-4 in your textbook. List two headings you would use to an outline of the lesson.

- _____
- _____

Vocabulary

Define *linear expression* in your own words..

_____
_____
_____

Real-World Link

Lacrosse Some of the statistics that are tracked in middle school lacrosse include goals and assists. The table shows the number of goals and assists that Jessica and Isabella scored in the first two games of the season.

Player	Number of Goals		Number of Assists	
	Game 1	Game 2	Game 1	Game 2
Jessica	g	3	2	1
Isabella	0	2	5	a

1. Write an expression to represent the total number of goals that each player scored in the first two games.

   Jessica: _____   Isabella: _____

2. Write an expression to show how many more goals Jessica scored than Isabella in the first two games. Then simplify the expression.

   _____

3. Write an expression to represent the total number of assists that each player had in the first two games.

   Jessica: _____   Isabella: _____

4. Write an expression to show how many more assists Isabella had than Jessica in the first two games. _____

5. Show the steps you would use to simplify the expression you wrote in Exercise 4. Justify each step.

   _____
   _____

Notes

Subtract Linear Expressions

For Exericses 1–3, circle the expression that is equivalent to the given expression.

1. $(5x + 2) - (3x + 1)$

$5x + 2 - 3x + 1$ $\qquad\qquad$ $5x + 2 - 3x - 1$

2. $(8c - 3) - (7c - 9)$

$8c - 3 - 7c + 9$ $\qquad\qquad$ $8c - 3 - 7c - 9$

3. $(4n + 5) - (2n + 6 - 5n)$

$4n + 5 - 2n - 6 + 5n$ $\qquad\qquad$ $4n + 5 - 2n - 6 - 5n$

4. What is one thing you want to remember about subtracting linear expressions?

_____

_____

Solve Problems with Linear Expressions

5. The expression $2m - 1$ represents the distance driven by the Nguyen family on Day 1 of a 3-day family vacation. The expression $5m + 6$ represents the total miles driven on the vacation. Write and simplify a subtraction expression that represents the miles driven on Days 2

and 3. _____

Summary

Write 2–3 sentences to summarize the lesson.

_____

_____

_____

_____

_____

_____

Rate Yourself!

Are you ready to move on? Shade the section that applies.

I have a few questions.

I'm ready to move on.

I have a lot of questions.

For more help, go online to access a Personal Tutor.

Tutor

 **FOLDABLES** Time to update your Foldable!

Factoring Linear Expressions

Getting Started

Scan Lesson 7-5 in your textbook. Predict two things you will learn about factoring linear expressions.

- _____
  _____

- _____
  _____

Vocabulary

Circle the vocabulary word defined below.

A number, a variable, or a product of a number and one or more variables.

factor   monomial

Real-World Link

Marching Band Band directors create geometrical formations that are eye-catching and exciting, but still stay with the rhythm and feel of the music. A band director is using a rectangular field that has an area of $(60x + 150)$ square yards. The director is separating the field into three equal-sized sections for the brass, woodwind, and percussion sections of the band.

1. Describe how you could find the area of the percussion section.

_____

_____

2. Find the area of the percussion section. Explain your reasoning.

_____

_____

_____

3. Suppose the field is 30 yards long, as shown below. Fill in the boxes to write a sentence that represents the area of the field.

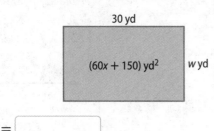

30 yd

$(60x + 150)$ yd^2 w yd

[ ] × [ ] = [ ]

4. What is the width of the field? _____

Notes

Find the GCF of Monomials

1. Cross out the linear expressions that do not belong.

$4x + 8$	$8t - 2$	$4m + 4n$
$20r - 18s$	$12v - 16$	$24c + 30$

2. What do all the remaining expressions in Exercise 1 have in common?

Factor Linear Expressions

3. Complete the graphic organizer to factor $15x + 10$.

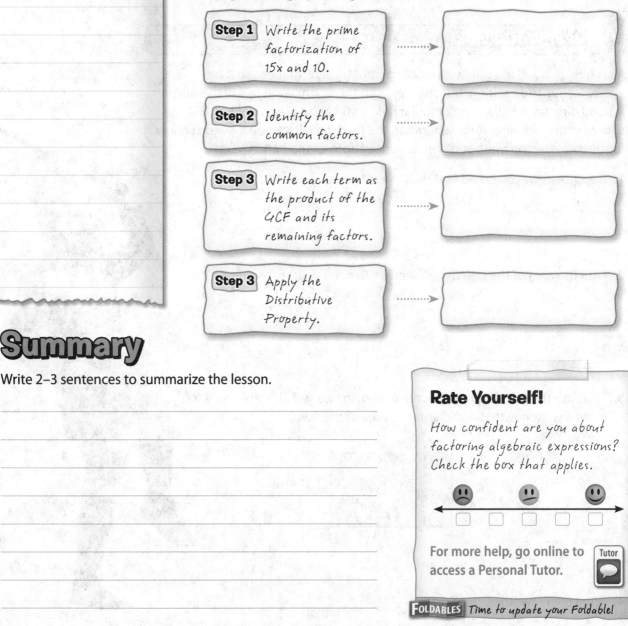

Step 1 Write the prime factorization of 15x and 10.

Step 2 Identify the common factors.

Step 3 Write each term as the product of the GCF and its remaining factors.

Step 3 Apply the Distributive Property.

Summary

Write 2–3 sentences to summarize the lesson.

Rate Yourself!

How confident are you about factoring algebraic expressions? Check the box that applies.

For more help, go online to access a Personal Tutor.

FOLDABLES Time to update your Foldable!

Chapter Review

Vocabulary Check

Reconstruct the vocabulary word and definition from the letters under the grid. The letters for each column are scrambled directly under that column.

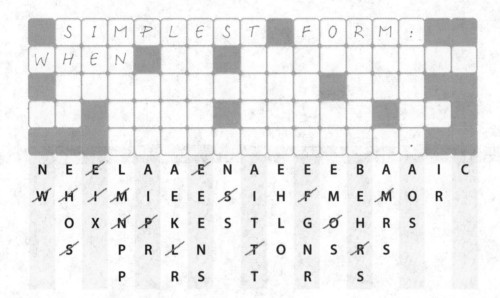

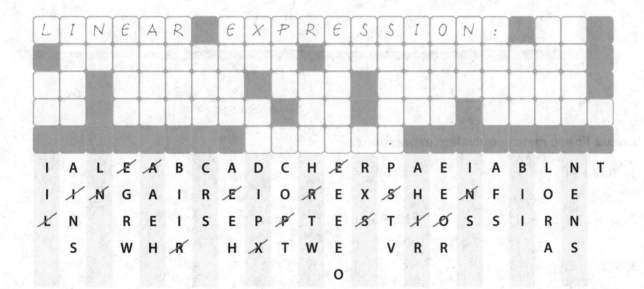

Use Your FOLDABLES

Use your Foldable to help review the chapter.

Tape here →

Linear Expressions

Explanation

Explanation

Got it?

Draw a line to match equivalent expressions.

1. $3(2x + 4)$ **a.** $6x + 4$

2. $(x + 2) + (5x + 2)$ **b.** $-6x + 9$

3. $4(2x + 5) - (2x + 3)$ **c.** $6x + 12$

4. $3x - (9x - 4) + 5$ **d.** $-6x - 9$

5. $-8x - (-2x + 9)$ **e.** $6x + 17$

Problem Solving

1. A music appreciation class is taking a field trip to the Rock and Roll Hall of Fame. The cost of admission and transportation for each of the 15 students is $20. Lunch will cost each person $8.00. Use mental math to find the total amount they will spend. Justify your answer by using the Distributive Property. (Lesson 1) _____

Show your work.

2. Write an expression in simplest form for the perimeter of the triangle. (Lesson 2) _____

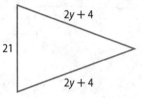

21

2y + 4

2y + 4

3. Pilar purchased a new computer. She made an initial payment of $70 and will make monthly payments of $45 for x months. Write an expression to show the total amount Pilar will pay for the computer. (Lesson 2)

4. A rectangle has side lengths (2x − 5) meters and (2x + 6) meters. Write and simplify an expression to represent the perimeter of the rectangle. (Lesson 3)

5. The cost for shipping a package that weighs x pounds is shown in the table. How much more does Mega Shipping charge than Delivery World? (Lesson 4) _____

Company	Cost ($)
Mega Shipping	3x + 3.50
Delivery World	2x + 2.99

6. Four friends visited the zoo to see the new shark exhibit. The group paid for admission to the zoo and $8 for parking. The total cost of the visit can be represented by the expression (4x + 8) dollars. What was the cost of the visit for one person? (Lesson 5) _____

Reflect

 **Answering the Essential Question**

Use what you learned about algebraic expressions to complete the graphic organizer. Explain why algebraic rules are useful. Then provide examples to illustrate your explanation.

Words

Essential Question

WHY are algebraic rules useful?

Examples

Answer the Essential Question. Why are algebraic rules useful?

Chapter 8
Equations and Inequalities

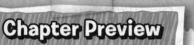

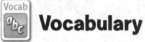 **Vocabulary**

empty set	identity	solution
equivalent equations	null set	two-step equation
inverse operations		

Vocabulary Activity

Complete the graphic organizer below.

Equations and Inequalities

Equation	Inequality
Describe It	Describe It
List Some Examples	List Some Examples

Example 1

Solve $x + 5 = 8$. Graph your solution on a number line.

$$x + 5 = 8$$
$$\underline{-5 = -5} \qquad \text{Subtract.}$$
$$x \quad = 3 \qquad \text{Simplify.}$$

To graph 3, draw a dot at 3 on the number line.

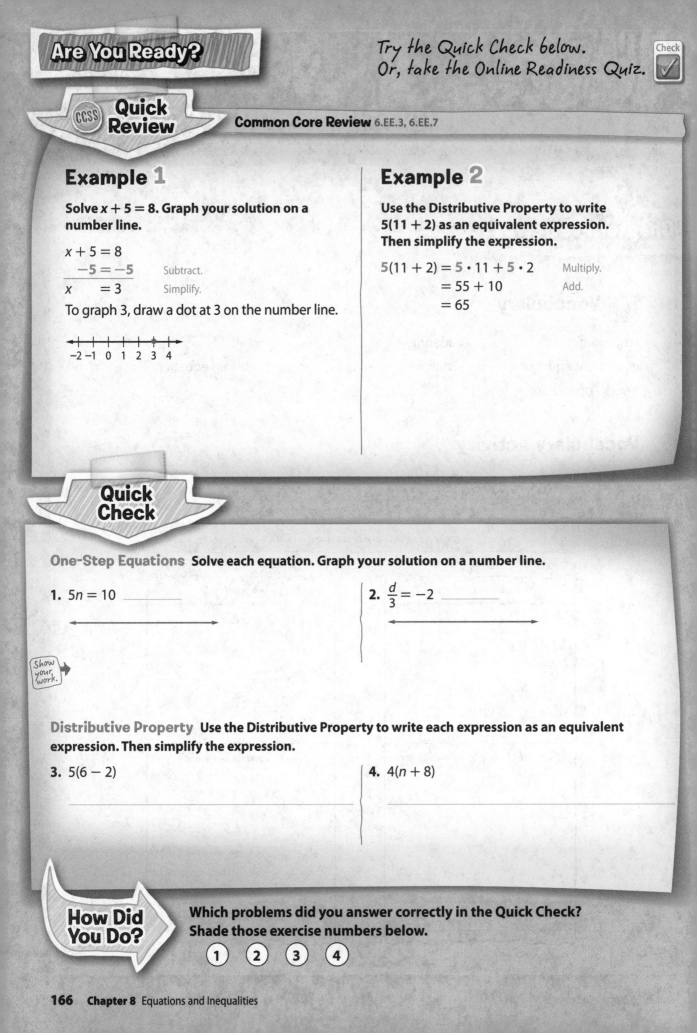

$$-2 \;-1 \;\; 0 \;\; 1 \;\; 2 \;\; 3 \;\; 4$$

Example 2

Use the Distributive Property to write $5(11 + 2)$ as an equivalent expression. Then simplify the expression.

$$5(11 + 2) = 5 \cdot 11 + 5 \cdot 2 \qquad \text{Multiply.}$$
$$= 55 + 10 \qquad \text{Add.}$$
$$= 65$$

Quick Check

One-Step Equations Solve each equation. Graph your solution on a number line.

1. $5n = 10$ _____

2. $\dfrac{d}{3} = -2$ _____

Show your work.

Distributive Property Use the Distributive Property to write each expression as an equivalent expression. Then simplify the expression.

3. $5(6 - 2)$

4. $4(n + 8)$

How Did You Do?

Which problems did you answer correctly in the Quick Check?
Shade those exercise numbers below.

① ② ③ ④

✂ cut on all dashed lines   ⬚ fold on all solid lines   ⬚ tape to page 188   **FOLDABLES**®

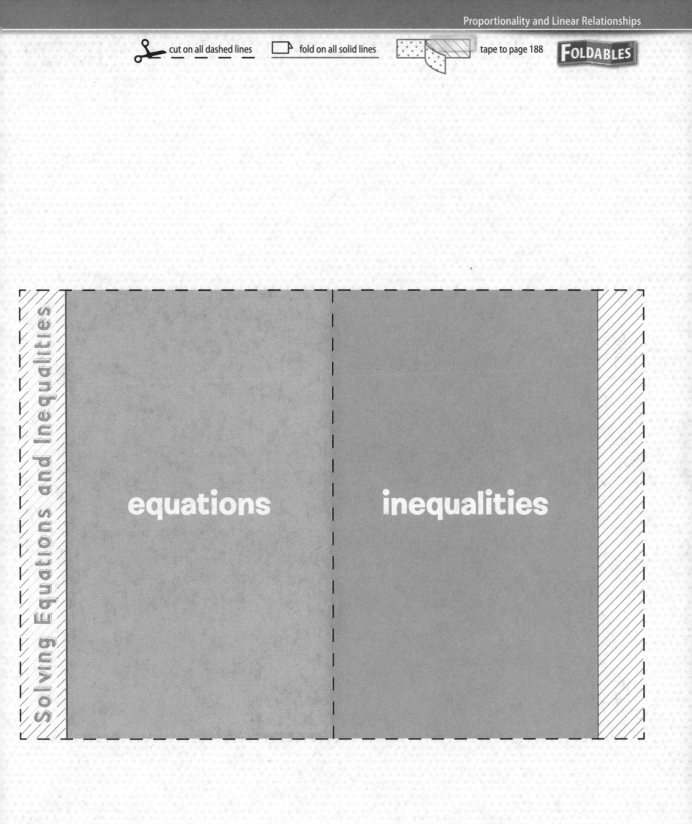

Solving Equations and Inequalities

equations

inequalities

FOLDABLES®
Study Organizer

1 Cut out the Foldable above.

2 Place your Foldable on page 188.

3 Use the Foldable throughout this chapter to help you solve equations and inequalities.

page 188

Two-Step Example(s)

Addition/Subtraction Example(s)

Multiplication/Division Examples

page 188

Tab 2

Tab 1

Solving Equations with Rational Coefficients

Getting Started

Scan Lesson 8-1 in your textbook. List two real-world scenarios in which you would solve equations with rational coefficients.

- _____
  _____
- _____
  _____

Vocabulary Start-Up

An **equation** such as $12 - 3 = 9$ or $3 + 2x = 21$, is a mathematical sentence that shows two expressions are equal. If an equation contains a variable, the value for the variable that makes the equation true is called a **solution**. For example, 9 is the solution of $3 + 2x = 21$ because $3 + 2(9) = 21$.

For each equation, underline the variable. Then circle the correct solution.

Equation	Possible Solutions		
$x + 0.5 = 17$	$x = 12$	$x = 16.5$	$x = 17.5$
$t - 10 = -20$	$t = -10$	$t = 10$	$t = -30$
$3n = 9.6$	$n = 3.2$	$n = 3.6$	$n = 28.2$
$\frac{d}{4} = 8$	$d = 2$	$d = \frac{1}{2}$	$d = 32$

Real-World Link

Social Networks More pre-teens and teens are participating in social networking than ever before. Three fourths of teens surveyed said they belong to a social network, compared to 40% of adults surveyed.

1. Suppose 750 teens surveyed said they belong to a social network. Let t represent the total number of teens surveyed and write an equation that models this situation. _____

2. Suppose 500 adults surveyed said they belong to a social network. Let a represent the total number of adults surveyed.

   Write an equation that models this situation. _____

Notes

Solve Equations by Dividing

1. Cross out the equation that does not belong.

$0.25t = 5$ $-2 = -0.1x$ $2.5 = -0.125s$

2. What is true about the remaining equations?

_____

Solve Equations by Multiplying

Complete the graphic organizer to solve $\frac{3}{4}c = 18$.

Step 1 Write the equation.	
Step 2 Use the Multiplication Property of Equality.	
Step 3 Write 18 as $\frac{18}{1}$. Divide by common factors.	
Step 4 Simplify.	

Summary

Write 2–3 sentences to summarize the lesson.

_____

_____

_____

_____

_____

_____

Rate Yourself!

☐ I understand how to solve equations with rational coefficients.

▶▶ Great! You're ready to move on!

☐ I still have questions about solving equations with rational coefficients.

▐▌ No Problem! Go online to access a Personal Tutor.

Solving Two-Step Equations

Getting Started

Scan Lesson 8-2 in your textbook. List two headings you would use to make an outline of the lesson.

- _____
  _____

- _____
  _____

Real-World Link

Cheerleading Cheerleaders on a middle school squad must purchase cheer shoes for $35, plus white ankle socks for $3 per pair. The total amount spent is $53. The equation $35 + 3x = 53$, where x is the number of pairs of socks, represents the total cost. You can use the *work backward* strategy to solve for x.

Start with the total cost. ☐

Subtract the cost of the shoes. − ☐

Since each pair of socks costs $3, divide by three. ☐ ÷ ☐ = ☐

The cheerleaders must buy ☐ pairs of ankle socks.

You can check your work by substituting your solution into the equation.

$35 + 3 \left(\boxed{} \right) \stackrel{?}{=} 53$

$35 + \boxed{} \stackrel{?}{=} 53$

$\boxed{} = 53$

1. How many pairs of socks are purchased if the total cost is $44?

   _____

2. The equation $15x + 90 = 135$ represents the total cost of x pom poms and the cheer uniform. How many sets of pom poms were purchased?

   _____

Notes

Solve Two-Step Equations

Evaluate the given solution for each equation. Mark an X through any incorrect solutions. Then find the correct solution.

1. $-3t + 8 = -4$ ·············> $t = 4$

2. $21 - \dfrac{y}{4} = 12$ ·············> $y = 16$

3. $7d - 15 = -71$ ·············> $d = -12$

4. $\dfrac{5s}{2} = -15$ ·············> $s = -6$

Solve Real-World Problems

5. Afyia wants to spend \$24 at an online music store. She buys one complete CD for \$6 and several single songs for \$2 each. Solve $24 = 2s + 6$ to find the number of single songs she can buy.

6. Alex went to the movies with several friends. Student tickets cost \$8.50 each, and together they spent \$25 on snacks. The total amount paid was \$59. Solve $8.5x + 25 = 59$ to find the number of people that went to the movies.

Summary

Write 2–3 sentences to summarize the lesson.

Rate Yourself!

How confident are you about solving two-step equations? Shade the ring on the target.

I'm on target.

I need help.

For more help, go online to access a Personal Tutor.

Tutor

FOLDABLES *Time to update your Foldable!*

Writing Equations

Getting Started

Scan Lesson 8-3 in your textbook. Predict two things you will learn about writing equations.

- _____
- _____
- _____

Quick Review

Translate each sentence into an equation.

Three times a number is twelve. _____

Twice a number is ten.

Real-World Link

Tablet Computers Accessories for tablet computers, such as docking stations, power adapters, and connection kits, help users get the most out of their tablets. Kara bought a case and a power adapter for her tablet. She paid $10 more for the case than the power adapter.

1. Which item cost more? _____

   How much more? _____

2. Suppose c represents the cost of the case. How much did the power adapter cost? _____

3. What expression could be used to represent the cost of the case and the power adapter? _____

4. Suppose Kara spent a total of $90 on both items. What equation could be used to represent the situation? _____

5. Use the equation from Exercise 4 to find the cost of each item.

   case: ☐

   power adapter: ☐

Notes

Write Two-Step Equations

Place the indicator words under the correct operation in the table.

twice   product   more than   decreased by
quotient   increased by   total   sum
less   times   difference   less than
into

Addition	Subtraction	Multiplication	Division

Two-Step Verbal Problems

Answer each question using the information below.

Together, Miguel and Carla spent $64 at the bookstore. Carla spent $15 less than Miguel.

1. Who spent less money? _____

2. How much less? _____

3. Write an expression to represent the amount of money Miguel spent, in terms of m. _____

4. Write an expression to represent the amount of money Carla spent, in terms of m. _____

5. Write an equation to represent the amount of money they spent altogether. _____

6. How much did each person spend at the bookstore?
   _____

Summary

Write 2–3 sentences to summarize the lesson.

_____

_____

_____

_____

_____

Rate Yourself!

Are you ready to move on? Shade the section that applies.

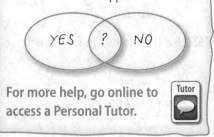

YES   ?   NO

For more help, go online to access a Personal Tutor.

More Two-Step Equations

Getting Started

Scan Lesson 8-4 in your textbook. List two headings you would use to make an outline of the lesson.

- _____

  _____

  _____

- _____

  _____

Real-World Link

Bowling Bowling alleys typically charge for the number of games played and the rental of bowling shoes. Kofi and two friends went bowling. Their total cost for games played and shoe rental was $48. Each person spent $2 to rent bowling shoes and paid the same amount of money for the games played.

1. Complete the bar diagram that represents the situation.

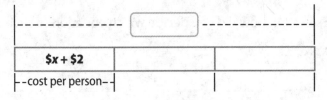

2. Use the bar diagram to complete the equation. $\boxed{}(x + 2) = \boxed{}$

3. From the diagram, you can see that Kofi spent one third of the total cost. So, Kofi spent $x + $2 = \dfrac{\$48}{3}$ or $\boxed{}$.

4. Suppose Kofi and three friends went bowling. If each person rented bowling shoes for $2 and their total cost was $56, write an equation that could represent this situation.

   _____

5. How could you use the equation you wrote in Exercise 4 to find the amount of money Kofi spent?

   _____

Notes

Solve Two-Step Equations

Complete the graphic organizer to solve 5(x − 2) = 22.

Step 1 Write the equation.	
Step 2 Use the Division Property of Equality.	
Step 3 Simplify.	
Step 4 Use the Addition Property of Equality.	
Step 5 Simplify.	

Use the Distributive Property

1. Mrs. Sanchez is making 5 costumes for the school play. Of the $60 she spent on material and supplies, Mrs. Sanchez spent $3 per costume for buttons and zippers. Circle the equation that represents this situation.

   $5x + 3 = 60$ $\qquad\qquad\qquad$ $5(x + 3) = 60$

2. In Exercise 1, how did you decide which equation to circle?

   _____

   _____

Summary

Write 2–3 sentences to summarize the lesson.

_____

_____

_____

_____

_____

_____

_____

Rate Yourself!

Are you ready to move on? Shade the section that applies.

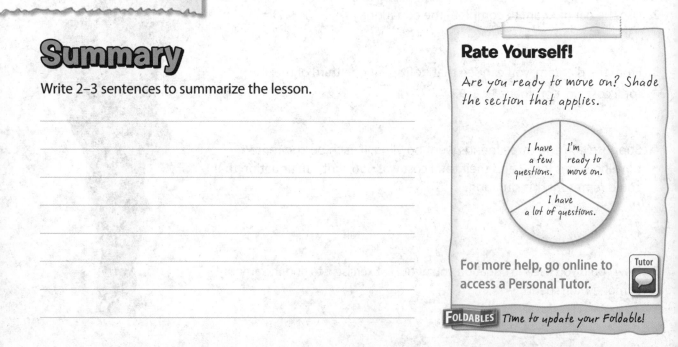

I have a few questions. / I'm ready to move on. / I have a lot of questions.

For more help, go online to access a Personal Tutor. **Tutor**

FOLDABLES Time to update your Foldable!

Mid-Chapter Check

Vocabulary Check

1. **Be Precise** Define *two-step equation*. Give an example of a two-step equation. Then solve your equation. (Lesson 2)

_____

_____

2. Describe how to solve the equation $\frac{2}{3}x = 6$. (Lesson 1)

_____

_____

Skills Check and Problem Solving

Solve each equation. Check your solution. (Lessons 1, 2, and 4)

3. $\frac{3}{4}x = -6$ _____

4. $\frac{5}{6}c + 7 = 17$ _____

5. $16 + 5w = 31$ _____

Show your work.

6. $28 = 4g - 4$ _____

7. $9(a - 4) = 27$ _____

8. $-5(3 + b) = 20$ _____

9. Mr. Carter is renting a car from an agency that charges $20 per day plus $0.15 per mile. He has a budget of $80 per day. Write and solve an equation to find the maximum number of miles he can drive each day. (Lesson 3)

_____

10. **Standardized Test Practice** Kenneth signed up to receive Internet service for $13 per month plus a $30 start-up fee. Which equation could be used to find the number of months he can receive Internet service for $134? (Lesson 3)

Ⓐ $134 + 30 = 13m$

Ⓒ $13 + 30m = 134$

Ⓑ $30 - 13m = 134$

Ⓓ $30 + 13m = 134$

21ST CENTURY CAREER
in Veterinary Medicine

Vet Techs Don't Monkey Around

For each problem, use the information in the tables to write an equation. Then solve the equation.

1. The minimum tail length of an emperor tamarin is 1.6 inches greater than that of a golden lion tamarin. What is the minimum tail length of a golden lion tamarin? _____

2. The minimum body length of a golden lion tamarin is 5.3 inches less than its maximum body length. What is the maximum body length? _____

3. Tamarins live an average of 15 years. This is 13 years less than the years that one tamarin in captivity lived. How long did the tamarin in captivity live? _____

4. The maximum weight of a golden lion tamarin is about 1.97 times the maximum weight of an emperor tamarin. What is the maximum weight of an emperor tamarin? Round to the nearest tenth. _____

5. For an emperor tamarin, the maximum total length, including the body and tail, is 27 inches. What is the maximum body length of an emperor tamarin?

_____

Emperor Tamarin Monkeys		
Measure	Minimum	Maximum
Body length	9.2 in.	b
Tail length	14 in.	16.6 in.
Weight	10.7 oz	w

Golden Lion Tamarin Monkeys		
Measure	Minimum	Maximum
Body length	7.9 in.	ℓ
Tail length	t	15.7 in.
Weight	12.7 oz	28 oz

Career Project

It's time to update your career portfolio! Go to the Occupational Outlook Handbook online and research a career as a veterinary technician. Include brief descriptions of the work environment, education and training requirements, and the job outlook.

_____

_____

_____

_____

Do you think you would enjoy a career as a veterinary technician? Why or why not?

- _____
- _____
- _____
- _____

Solving Equations with Variables on Each Side

Getting Started

Scan Lesson 8-5 in your textbook. List two real-world scenarios in which you would solve equations with variables on each side.

- _____
  _____

- _____

Vocabulary

Write the definition of *Addition Property of Equality* in your own words.

_____

_____

_____

Real-World Link

Camping Many campsites offer rentals for equipment, like kayaks and bicycles. The table shows the rental fees for a certain campground.

Item	Deposit ($)	Cost per Day ($)
Bicycle	3.00	5.50
Kayak	6.00	5.00

1. What expressions could be used to find the total cost of renting each item for any number of days *d*?

   bicycle: _____   kayak: _____

2. Use a table to find the number of days that you would need to rent each item for the costs to be the same. The first one is done for you.

Days	Bicycle Cost ($)	Kayak Cost ($)
1	8.50	11.00
2		
3		
4		
5		
6		
7		

3. When does it cost less to rent a bicycle? _____

4. When does it cost less to rent a kayak? _____

5. When does it cost the same? _____

Notes

Equations with Variables on Each Side

The equation $3x - 4 = 5x - 2$ is modeled below. Explain how you would use algebra tiles to solve the equation.

$$3x - 4 \qquad = \qquad 5x - 2$$

Solve Verbal Problems

1. A wireless company offers two cell phone plans. Plan A charges $10 per month plus $0.25 per text message. Plan B charges $25 per month plus $0.15 per text message. Write and solve an equation to find the number of text messages that would make the plans cost the same.

Summary

Write 2–3 sentences to summarize the lesson.

Rate Yourself!

How confident are you about solving equations with variables on each side? Check the box that applies.

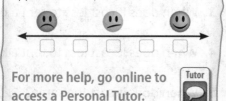

For more help, go online to access a Personal Tutor.

Getting Started

Write the math and the real-world definitions of inequality.

- math definition: _____

  _____

- real-world definition: _____

  _____

Quick Review

What is the meaning of each symbol?

> _____

< _____

🌎 Real-World Link

Water Parks Wisconsin Dells, Wisconsin, is known as the Water Park Capital of the World. The town has 20 water parks with more than 200 waterslides and 16 million gallons of water. The table shows the admission rates for one of the parks.

Type of Ticket	Price ($)
Child 47 inches tall and under	25
Adult Over 47 inches tall	35

1. What is the height requirement to purchase an adult ticket?

   _____

2. What is the maximum height of a person that can purchase a child ticket?

   _____

3. The Blackfox family is going to the park. Circle the type of ticket Mr. Blackfox needs to buy for each family member.

Mr. Blackfox	72 inches tall	child	adult
Mrs. Blackfox	64 inches tall	child	adult
Lupe	42 inches tall	child	adult
Juan	47 inches tall	child	adult
Rosa	58 inches tall	child	adult

4. What type of ticket did he buy for Juan? Explain.

   _____

   _____

5. How tall are you? Would you need to buy an adult ticket? Explain.

   _____

Notes

Write Inequalities

Write $<$, $>$, $\leq$, or $\geq$ to represent each phrase. The first one has been done for you.

Inequalities	
Phrase	Symbol
is greater than	>
is no more than	
is at least	
is fewer than	
exceeds	
is no less than	
is at most	

Graph Inequalities

For each inequality, write *closed* or *open* to indicate which type of circle you would use to graph the inequality on a number line. Then indicate whether the arrow would point *right* or *left*.

1. $x \geq -5$ _____

2. $x < 12$ _____

3. $-8 > x$ _____

4. $x \leq 4$ _____

5. $x < -6$ _____

6. $3 \leq x$ _____

Summary

Write 2–3 sentences to summarize the lesson.

_____

_____

_____

_____

_____

_____

_____

Rate Yourself!

How well do you understand writing and graphing inequalities? Circle the image that applies.

Clear   Somewhat Clear   Not So Clear

For more help, go online to access a Personal Tutor.

Tutor

Solving Inequalities

Getting Started

Scan Lesson 8-7 in your textbook. List two headings you would use to make an outline of the lesson.

- _____
  _____
  _____

- _____
  _____
  _____

Vocabulary
Write the definition of *inequality* in your own words.

_____
_____
_____
_____

Real-World Link

Pets Did you know that 39% of U.S. households own at least one dog? The amount of food that you feed your dog should be based on the dog's weight. Jackson has a Labrador retriever that weighs 65 pounds and should eat no more than $2\frac{1}{2}$ cups of dog food each day.

1. Which inequality symbol would you use to represent the phrase

   no more than? ☐

2. Suppose Jackson feeds his dog twice each day. If *a* represents the amount of feed he gives the dog at each feeding, what is the meaning of the inequality below?

   $$2a \leq 2\frac{1}{2}$$

   _____
   _____

3. Rewrite the inequality by replacing the $\leq$ sign with $=$. _____

   How would you solve this equation? _____

   So, $a = $ ☐

4. Replace the equals sign with the less than or equal to symbol.

   $$a \leq \boxed{}$$

   What is the meaning of this new inequality?

   _____
   _____

Notes

Addition and Subtraction Properties

Complete the graphic organizer by writing the steps to solve the inequality.

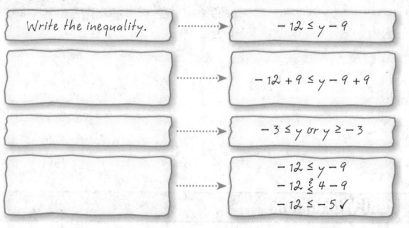

| Write the inequality. | ⟩ | $-12 \le y - 9$ |

| | ⟩ | $-12 + 9 \le y - 9 + 9$ |

| | ⟩ | $-3 \le y \text{ or } y \ge -3$ |

| | ⟩ | $-12 \le y - 9$
 $-12 \overset{?}{\le} 4 - 9$
 $-12 \le -5 \checkmark$ |

Multiplication and Division Properties

Solve each inequality. Then draw a line to match the solution to its corresponding number line.

1. $\dfrac{z}{6} > 4$

   a. ◄ 17 18 19 20 21 22 23 24 25 26 27 ►

2. $-3n \ge -60$

   b. ◄ 17 18 19 20 21 22 23 24 25 26 27 ►

3. $-2g \le -44$

   c. ◄ 17 18 19 20 21 22 23 24 25 26 27 ►

Summary

Write 2–3 sentences to summarize the lesson.

_____

_____

_____

_____

_____

_____

Rate Yourself!

Are you ready to move on? Shade the section that applies.

I have a few questions.

I'm ready to move on.

I have a lot of questions.

For more help, go online to access a Personal Tutor.

Tutor

FOLDABLES Time to update your Foldable!

Solving Multi-Step Equations and Inequalities

Getting Started

Scan Lesson 8-8 in your textbook. Write the definitions of identity and null set.

- identity: _____

  _____

- null set: _____

  _____

Real-World Link

Field Trip Mr. Murphy's class of 20 students is going on a field trip to the science center. Admission to the museum is $8 per student and there is an additional cost of *m* dollars per student to watch the 3-D movie. The total cost for all of the students is $270.

1. Fill in the information that you know.

   cost of admission per student ☐

   cost of movie per student ☐

   number of students ☐

   total cost for all students ☐

2. What expression can be used to represent the total cost per student?

   _____

3. What expression can be used to represent the total cost of admission and a movie for all students?

   _____

4. Use the Distributive Property to rewrite the expression from Exercise 3 as an equivalent expression.

   _____

5. Using the expression for Exercise 4, write and solve an equation to find the cost of a ticket for the 3-D movie.

   _____

Notes

Solve Multi-Step Equations

Write an equation that has a solution that is an identity. Then write an equation that has a solution that is the empty set.

identity: _____

null or empty set: _____

Solve Multi-Step Inequalities

Complete the steps in the table to solve $-4(x + 12) > -(3x + 16)$.

Solve Multi-Step Inequalities	
Step 1 Write the equation.	
Step 2 Use the Distributive Property on both sides of the equation.	
Step 3 Add 3x to both sides of the equation.	
Step 4 Add 48 to both sides of the equation.	
Step 5 Multiply both sides of the equation by -1. Reverse the inequality symbol.	

Summary

Write 2–3 sentences to summarize the lesson.

_____

_____

_____

_____

_____

_____

Rate Yourself!

☐ I understand how to solve multi-step equations and inequalities.

▶▶ Great! You're ready to move on!

☐ I still have questions about solving multi-step equations and inequalities.

No Problem! Go online to access a Personal Tutor.

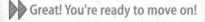

FOLDABLES Time to update your Foldable!

Chapter Review

Vocabulary Check

Fill in the blank with the correct vocabulary term. Then circle the word that completes the sentence in the word search.

1. The _____ or _____ is shown by the symbol Ø.

2. A(n) _____ is a mathematical sentence that contains a less than or greater than symbol.

3. The inequality symbol must be reversed when you multiply or divide both sides by a _____ number.

4. The inequality symbol remains the same when you multiply or divide both sides by a _____ number.

5. In the equation $\frac{3}{4}x + 4 = 12$, $\frac{3}{4}$ is a _____ coefficient.

6. A value for the variable that makes an equation true is called a(n) _____.

7. An equation that contains two steps is called a(n) _____.

8. A(n) _____ is a mathematical sentence that contains an equals sign.

9. A(n) _____ is an equation that is true for every value of the variable.

Q	C	J	P	X	L	G	P	Y	F	S	A	F	W	F	W	T	I	Q	A	M	I	Q	P	W
E	V	I	T	A	G	E	N	X	T	C	K	O	S	A	K	E	O	G	R	X	K	O	D	F
J	F	G	E	O	H	J	F	N	Q	I	T	X	H	N	Z	S	A	O	C	O	S	K	H	F
X	X	H	B	G	S	U	A	F	F	I	L	V	A	T	U	Y	E	V	G	I	E	P	U	B
D	D	S	W	W	N	H	Z	K	M	G	C	A	N	Y	L	T	A	P	T	F	P	F	V	C
N	Q	G	O	B	V	E	R	H	F	E	M	E	U	D	B	P	P	I	B	I	D	G	R	H
I	C	G	O	L	U	N	Z	B	Y	X	A	Q	P	Q	D	M	V	O	N	O	M	U	D	W
N	O	I	T	A	U	Q	E	P	E	T	S	O	W	T	E	E	L	O	A	N	M	H	W	D
L	U	D	S	A	G	T	K	S	G	A	O	L	M	K	G	N	I	A	W	C	S	V	V	F
O	B	E	A	K	Z	I	I	U	K	Q	A	K	L	M	Z	T	I	U	Z	I	W	D	E	J
A	R	N	U	H	V	V	W	O	O	F	F	G	L	U	A	I	O	G	R	Z	Y	O	A	R
R	W	T	J	V	R	Y	N	P	N	H	U	E	K	U	N	C	L	A	N	O	I	T	A	R
F	L	I	B	N	S	N	L	M	W	G	D	C	Q	E	G	N	V	K	H	F	L	T	A	C
V	V	T	P	F	D	I	G	S	V	F	I	E	W	I	R	Z	V	U	P	N	I	E	A	V
G	E	Y	E	J	W	J	F	D	Z	W	I	C	H	H	B	G	H	L	V	R	W	J	E	P

Use Your FOLDABLES

Use your Foldable to help review the chapter.

Tape here

Tab 1 **Solving Equations and Inequalities**

Multi-Step Example(s)

Multi-Step Example(s)

Tape here

Tab 2

Got it?

Number the steps in the order needed to solve each equation. Then solve the equation.

1. $3(x + 6) = -18$

_____ Subtract 18 from each side.

_____ Divide each side by 3.

_____ Multiply x and 6 by 3.

$x = $ _____

2. $4x - 7 = 6x - 5$

_____ Divide each side by 2.

_____ Subtract $4x$ from each side.

_____ Add 5 to each side.

$x = $ _____

3. $\frac{1}{3}(x - 12) = \frac{2}{3}x - 6$

_____ Multiply each side by 3.

_____ Multiply x and 12 by $\frac{1}{3}$.

_____ Add 6 to each side.

_____ Subtract $\frac{1}{3}x$ from each side.

$x = $ _____

Problem Solving

1. An online music company advertises the rates shown in the table. Sherita has $30 to pay the membership fee and download songs. (Lessons 1 and 2)

Type of Fee	Cost ($)
Membership	$8.75
Song Download	$0.85

   a. Solve the equation $0.85s + 8.75 = 30$ to find the number of songs that she can download. _____

   Show your work.

   b. If the membership fee increases to $11.30, how many songs can she download? _____

2. The Trans-Pacific Express project is an 11,000-mile long telecommunications cable under the Pacific Ocean connecting the United States to eastern Asia. The cost of the project was about $500 million. Write and solve an equation to find the cost per mile to the nearest dollar. (Lesson 3)

3. Theresa and Miranda are each saving money for a cruise. Theresa has already saved $500 and plans to deposit $40 each month. Miranda has $200 in her account and will deposit $60 each month. Write and solve an equation to find how many months it will take for them to have saved the same amount of money. (Lesson 5)

4. Mt. Waialeale in Hawaii receives an average rainfall of at least 397 inches per year. Write an inequality to describe the amount of rainfall. (Lesson 6) _____

5. Jin earns $2,350 per month plus $45 for each sale he completes. Write and solve an inequality to find how many sales he would have to make each month in order to earn at least $3,000. (Lesson 7) _____

6. Mimi is planning to run a marathon. To prepare for the race, she will follow the schedule below. She plans on running 11 hours per week. How many hours will she run each day? (Lesson 8)

Running Schedule	
Day	**Length of Time**
Monday	x hours
Tuesday	2 hours more than Monday
Thursday	same as Monday
Saturday	3 times as much as Monday

Reflect

 Answering the Essential Question

Use what you learned about equations and inequalities to complete the graphic organizer.

When do you use an equals sign?

 Essential Question

HOW are equations and inequalities used to describe and solve multi-step problems?

When do you use an inequality symbol?

 Answer the Essential Question HOW are equations and inequalities used to describe and solve multi-step problems?

Chapter Preview

Vocabulary

constant of variation	dependent variable	slope-intercept form
constant rate of change	independent variable	substitution
direct variation	linear equation	system of equations
function	linear relationship	vertical line test
function notation	rate of change	x-intercept
function rule	slope	y-intercept
function table		

Vocabulary Activity

Use the glossary to find the definition of *function notation*. Then label the figure with the terms *function notation*, *dependent variable*, and *independent variable*.

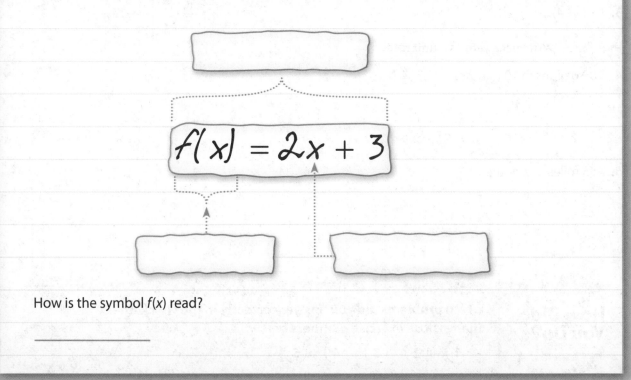

$$f(x) = 2x + 3$$

How is the symbol $f(x)$ read?

_____

Are You Ready?

*Try the Quick Check below.
Or, take the Online Readiness Quiz.*

Check ✓

CCSS Quick Review   **Common Core Review** 6.NS.6, 7.RP.1

Example 1

Use the coordinate plane to name the point for (−1, 2).

Step 1 Start at the origin, (0, 0).

Step 2 Move 1 unit left.

Step 3 Move 2 units up.

So, point *K* is at (−1, 2).

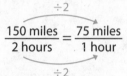

Example 2

Write 150 miles in 2 hours as a unit rate.

Write the rate that compares the number of miles to the number of hours.

$$\frac{150 \text{ miles}}{2 \text{ hours}} = \frac{75 \text{ miles}}{1 \text{ hour}}$$
÷2 ... ÷2

So, the unit rate is 75 miles per hour.

Quick Check

Coordinate Plane Use the coordinate plane to name the point for each ordered pair.

1. (3, −4) _____

2. (0, 4) _____

3. (−1, −4) _____

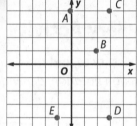

Unit Rate Write each rate as a unit rate.

4. 180 gallons in 10 minutes _____

Show your work.

5. 455 miles in 7 hours _____

How Did You Do?   **Which problems did you answer correctly in the Quick Check?
Shade those exercise numbers below.**

① ② ③ ④ ⑤

✂ cut on all dashed lines   ⬚ fold on all solid lines   ▨ tape to page 212   **FOLDABLES**

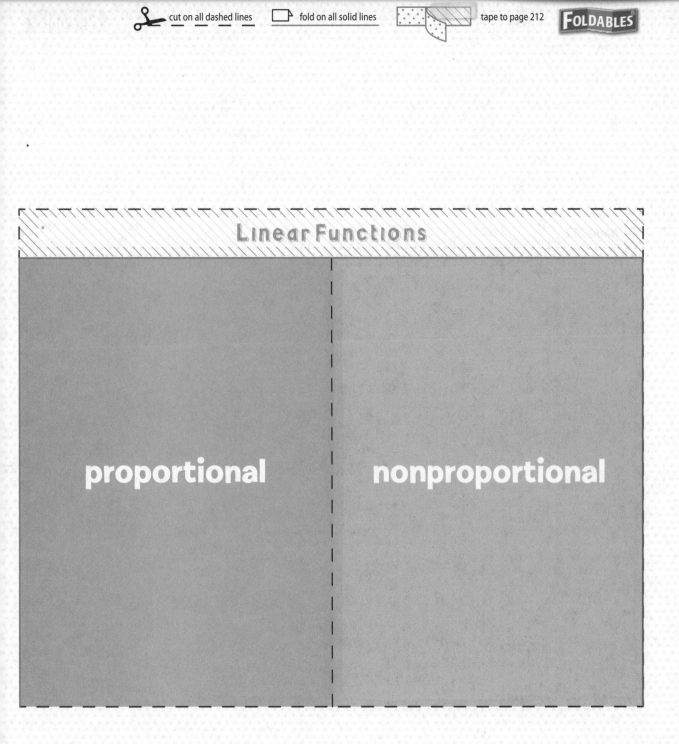

Linear Functions

proportional

nonproportional

FOLDABLES
Study Organizer

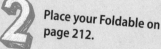

1 Cut out the Foldable above.

2 Place your Foldable on page 212.

3 Use the Foldable throughout this chapter to help you learn about linear functions.

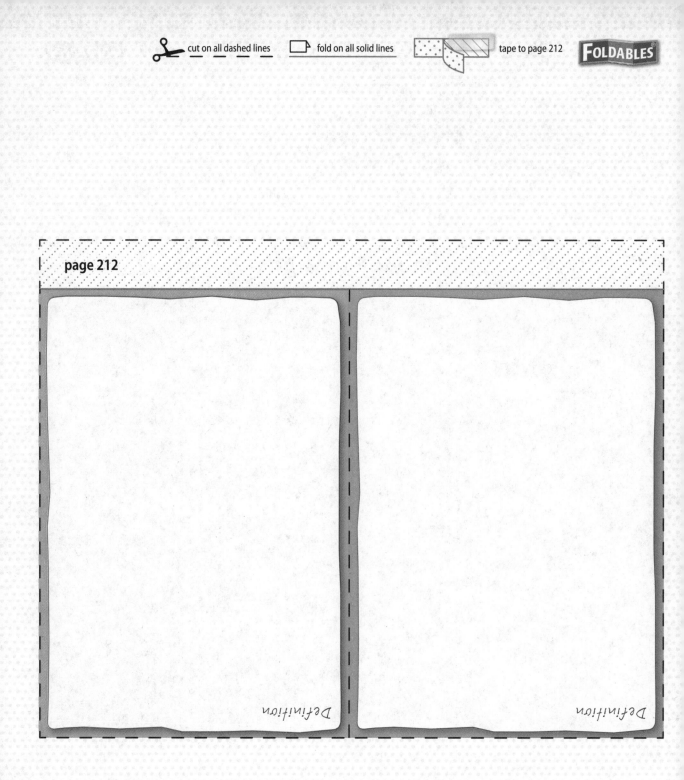

page 212

Definition

Definition

Functions

Getting Started

Scan Lesson 9-1 in your textbook. List two headings you would use to make an outline of the lesson.

- _____
  _____
- _____
  _____

Vocabulary

Ⓒircle the vocabulary word defined below.

The set of all *x*-coordinates from a set of ordered pairs.

domain   range

Vocabulary Start-Up

A set of ordered pairs such as {(2, 3), (3, 5), (4, 1), (5, 6)} is a relation. A relation is a **function** when each domain value is paired with exactly one range value.

Complete the graphic organizer below.

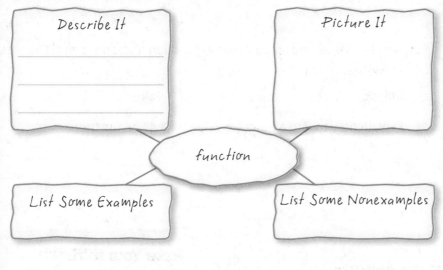

Describe It

Picture It

function

List Some Examples

List Some Nonexamples

🌐 Real-World Link

Meerkats Meerkats have sharp claws to forage for food and dig underground burrows. A typical meerkat burrows has an average of 15 entrance and exit holes.

1. For each ordered pair in the relation below, the *x*-coordinate represents the number of burrows and the *y*-coordinate represents the number of entrance and exit holes. Complete the relation if each burrow has 15 exit and entrance holes.

   {(1, ☐), (2, ☐), (3, ☐), (☐ , 60)}

2. Write a rule that represents the relation in Exercise 1. _____

Notes

Relations and Functions

1. Complete the graphic organizer on relations and functions.

Relations and Functions

A relation written as a set of ordered pairs is a function when

_____

_____

A relation given as a graph in a coordinate plane is a function when

_____

_____

Describe Relationships

Keisha spent $1.99 for each app she downloaded.

2. Write a function to find the cost of n number of downloaded apps.

_____

3. Write and evaluate an equation in function notation to find the cost of 5 apps.

function: _____   value: _____

4. Explain how to find a function value for a given function.

_____

_____

Summary

Write 2–3 sentences to summarize the lesson.

_____
_____
_____
_____
_____
_____
_____

Rate Yourself!

Are you ready to move on? Shade the section that applies.

I have a few questions.

I'm ready to move on.

I have a lot of questions.

For more help, go online to access a Personal Tutor.

Tutor

Representing Linear Functions

Getting Started

Scan Lesson 9-2 in your textbook. Predict two things you will learn about representing linear functions.

- _____
  _____
  _____

- _____
  _____
  _____

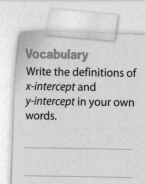

Vocabulary

Write the definitions of *x-intercept* and *y-intercept* in your own words.

_____

_____

_____

_____

Real-World Link

Racing The Daytona 500 was first run in 1959 and the average speed of the winner was about 135 miles per hour. The record for the fastest average speed is 177.6 miles per hour, which is about 3 miles per minute.

1. Complete the table to find the distance traveled at the record speed after 5, 10, 15, and 20 minutes.

Total Distance Traveled		
Time in Minutes	**3x**	**Distance in Miles**
5	3(5)	15
10		
15		
20		

2. Graph the ordered pairs (time, distance). Then connect the points.

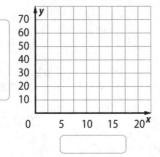

Total Distance Traveled

3. Do the data represent a function? Explain. _____

_____

4. Write an equation representing the relationship between time *x* and

distance *y*. _____

Notes

Solve Linear Functions

Complete each table. Then write the ordered pairs under the table.

1. $y = 3x + 1$

x	3x + 1	y
−2	3(−2) + 1	−5
		1
1		
		7

2. $y = -x + 2$

x	−x + 2	y
−1	−(−1) + 2	3
0		
	−(2) + 2	0
3		

Graph Linear Functions

Complete the graphic organizer that compares the two methods of graphing a linear function.

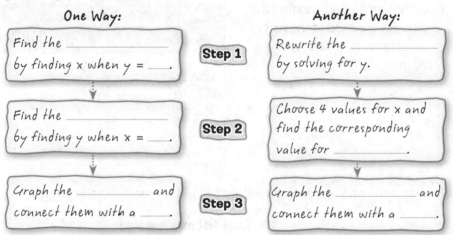

One Way:

Find the _____ by finding x when y = ____.

Step 1

Find the _____ by finding y when x = ____.

Step 2

Graph the _____ and connect them with a ____.

Step 3

Another Way:

Rewrite the _____ by solving for y.

Choose 4 values for x and find the corresponding value for _____.

Graph the _____ and connect them with a ____.

Summary

Write 2–3 sentences to summarize the lesson.

_____
_____
_____
_____
_____
_____

Rate Yourself!

Are you ready to move on? Shade the section that applies.

YES   ?   NO

For more help, go online to access a Personal Tutor.

Tutor

Constant Rate of Change and Slope

Getting Started

Scan Lesson 9-3 in your textbook. Write the definitions of rate of change and linear relationship.

- rate of change _____

  _____

- linear relationship _____

Quick Review
Simplify $\frac{20 - 12}{15 - 13}$.

_____

Vocabulary Start-Up

In a linear relationship, the rate of change is the same, or constant.

Complete the graphic organizer below.

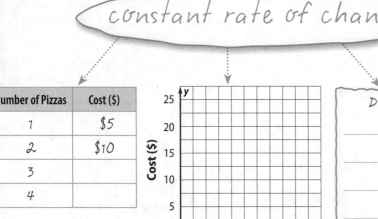

constant rate of change

Number of Pizzas	Cost ($)
1	$5
2	$10
3	
4	

Describe It

Real-World Link

Exercise Routines Treadmills, elliptical trainers, and free weights are among the most popular pieces of exercise equipment. The incline of a treadmill can determine the number of calories burned. Circle the correct statement.

The steeper the incline of the treadmill, the fewer calories burned.	The steeper the incline of the treadmill, the more calories burned.

Notes

Constant Rate of Change

1. Cross out the set of coordinates in the circle that do not belong. Then describe what the remaining sets have in common.

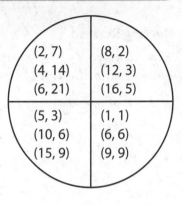

(2, 7)	(8, 2)
(4, 14)	(12, 3)
(6, 21)	(16, 5)
(5, 3)	(1, 1)
(10, 6)	(6, 6)
(15, 9)	(9, 9)

_____

_____

_____

_____

Slope

2. Complete the table on how to find slope.

Using a Graph	Using Coordinates of Points

Summary

Write 2–3 sentences to summarize the lesson.

_____

_____

_____

_____

_____

_____

_____

Rate Yourself!

How confident are you about constant rate of change? Shade the ring on the target.

I'm on target.

I need help.

For more help, go online to access a Personal Tutor.

Tutor

FOLDABLES *Time to update your Foldable!*

Direct Variation

Getting Started

Scan Lesson 9-4 in your textbook. List two heading your would use to make an outline of the lesson.

- _____
  _____
- _____
  _____

Vocabulary

Circle the type of relationship in which the ratio of two quantities is constant.

proportional

nonproportional

Real-World Link

Video Games According to a recent survey, about 56% of U.S. households own a current video game system. Games for these systems can get expensive, so many stores offer sales on pre-owned video games. The prices for pre-owned games at a local store are shown in the table below.

Number of Games	Total Cost ($)
2	16.50
4	33.00
6	49.50
8	66.00

Recall that when the ratio of two variable quantities is constant, a proportional relationship exists. This relationship is called a *direct variation*. The constant ratio is called the *constant of variation* or *constant of proportionality*.

1. Complete the steps below to derive the equation for a direct variation.

$$\frac{\boxed{}}{\boxed{}} = \boxed{}$$ Slope formula

$$\frac{y - 0}{x - 0} = m$$ $(x_1, y_1) = (0, 0)$
$(x_2, y_2) = (x, y)$

$$\frac{\boxed{}}{\boxed{}} = m$$ Simplify.

$$y = \boxed{}\boxed{}$$ Multiplication Property of Equality

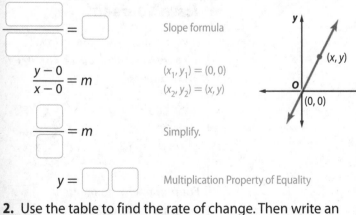

2. Use the table to find the rate of change. Then write an equation in $y = mx$ form to represent the situation.

_____

Questions

Notes

Direct Variation

Complete the graphic organizer.

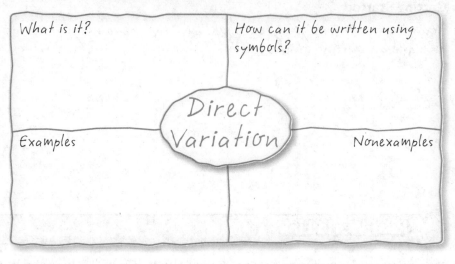

What is it?

How can it be written using symbols?

Direct Variation

Examples

Nonexamples

Compare Direct Variations

The distance that Car A travels varies directly with the number of hours, as shown in the table at the right. Car B's distance traveled can be represented by the equation $y = 50x$. Which car travels faster? Explain.

Time (h), x	Distance (mi), y
2	110
4	220
6	330
8	440

Summary

Write 2–3 sentences to summarize the lesson.

Rate Yourself!

How well do you understand direct variation? Circle the image that applies.

Clear   Somewhat Clear   Not So Clear

For more help, go online to access a Personal Tutor.

Tutor

FOLDABLES _Time to update your Foldable!_

202 Chapter 9 Linear Functions

Mid-Chapter Check

Vocabulary Check

Vocab
abc

1. **CCSS Be Precise** Define *function*. Give an example of a function. (Lesson 1)

_____

_____

2. Explain why the equation $y = 2x - 4$ is not a direct variation. (Lesson 4)

Skills Check and Problem Solving

Determine whether each relation is a function. Explain. (Lesson 1)

3. $\{(0, 5), (1, 2), (1, -3), (2, 4)\}$

_____

_____

4. $\{(-6, 3), (-3, 4), (0, 4), (3, 5)\}$

_____

_____

Graph each function. (Lesson 2)

5. $y = x + 2$

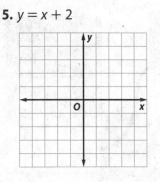

6. $y = 3x - 4$

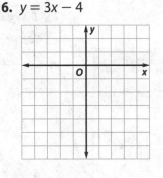

7. $y = -2x + 3$

8. Find the constant rate of change for the linear function shown at the right and interpret its meaning. (Lesson 3)

_____

9. What is the slope of the line that passes through the points $R(11, 6)$ and $S(9, -1)$? (Lesson 3) _____

10. **Standardized Test Practice** The total profit for a school fundraiser varies directly with the number of potted plants sold. The school earns $57.60 if 12 plants are sold. Which equation could be used to find the profit per plant sold? (Lesson 3)

Ⓐ $y = 57.6x$ Ⓑ $y = 12x$ Ⓒ $y = 4.8x$ Ⓓ $y = x$

Selling T-shirts

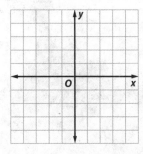

21ST CENTURY CAREER in Music

Mastering the Music

Use the information in the tables to solve each problem.

1. Is the set of ordered pairs (number of songs, cost) in each table a function? Explain.

   _____

   _____

2. Identify the independent and dependent variables for Dynamic Mastering. Then write a function to represent the total cost of any number of songs. _____

   _____

3. Is there a proportional linear relationship between number of songs and cost at Dynamic Mastering? Explain. _____

   _____

4. Write a direct variation equation to represent the number of songs *x* and cost *y* at Dynamic Mastering. How much does it cost to master 11 songs? _____

5. Find the slope of the line represented in the Mastering Mix table. What does the slope represent? _____

   _____

6. Is the linear relationship represented in the Engineering Hits table a direct variation? Explain. _____

   _____

Engineering Hits

Number of Songs	Cost ($)
1	100
2	160
3	210
4	250

Dynamic Mastering

Number of Songs	Cost ($)
2	120
4	240
6	360
8	480

Mastering Mix

Number of Songs	Cost ($)
1	125
3	275
5	425
7	575

Career Project

It's time to update your career portfolio! Find the name of the mastering engineer on one of your CDs. Use the Internet or another source to write a short biography of this engineer. Include a list of other artists whose songs he or she has mastered.

_____

_____

_____

_____

Do you think you would enjoy a career as a mastering engineer? Why or why not?

_____

_____

_____

Lesson 9-5
Slope-Intercept Form

Getting Started

Scan Lesson 9-5 in your textbook. Predict two things you will learn about the slope-intercept form.

- _____
  _____
  _____

- _____
  _____
  _____

Vocabulary
Write the definition of *linear relationship* in your own words.

_____

_____

_____

_____

Real-World Link

Space Camp A week-long space camp costs $800. Marissa's parents paid an initial $400 deposit, and then pay the rest in monthly payments of $100, as shown in the table.

In a nonproportional linear relationship, the graph passes through the point $(0, b)$ or the y-intercept.

Number of Months	Total Amount Paid ($)
0	400
1	500
2	600
3	700

1. Complete the steps to derive the equation for a nonproportional linear relationship by using the slope formula.

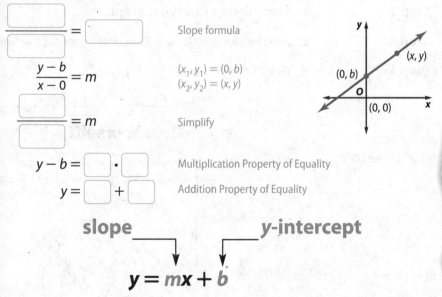

$$\frac{\boxed{}}{\boxed{}} = \boxed{} \qquad \text{Slope formula}$$

$$\frac{y - b}{x - 0} = m \qquad \begin{array}{l}(x_1, y_1) = (0, b)\\(x_2, y_2) = (x, y)\end{array}$$

$$\frac{\boxed{}}{\boxed{}} = m \qquad \text{Simplify}$$

$$y - b = \boxed{} \cdot \boxed{} \qquad \text{Multiplication Property of Equality}$$

$$y = \boxed{} + \boxed{} \qquad \text{Addition Property of Equality}$$

slope **y-intercept**

$$y = mx + b$$

2. Is the relationship between number of months and total amount paid proportional? Explain.

_____

_____

Notes

Slope-Intercept Form

Complete the graphic organizer.

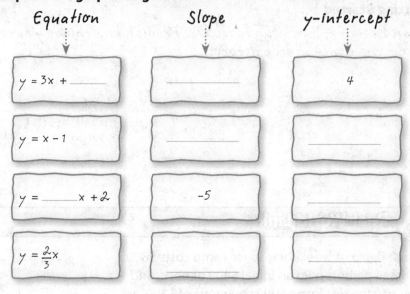

Equation   Slope   y-intercept

Equation	Slope	y-intercept
$y = 3x +$ _____	_____	4
$y = x - 1$	_____	_____
$y =$ _____ $x + 2$	-5	_____
$y = \frac{2}{3}x$	_____	_____

Graph Equations

Draw lines to correctly show the steps for graphing an equation using the slope-intercept form.

Step 1   a. Draw a line through the two points.

Step 2   b. Find the slope and *y*-intercept.

Step 3   c. Use the slope to locate a second point on the line.

Step 4   d. Graph the *y*-intercept point.

Summary

Write 2–3 sentences to summarize the lesson.

_____

_____

_____

_____

_____

_____

Rate Yourself!

How confident are you about slope-intercept form? Shade the ring on the target.

I'm on target.

I need help.

For more help, go online to access a Personal Tutor.

Tutor

Solve Systems of Equations by Graphing

Getting Started

Scan Lesson 9-6 in your textbook. Write the math and the real-world definitions of system.

- math definition _____

  _____

- real-world definition _____

Quick Review
Explain how to graph two points on the line $y = 3x + 2$.

_____

_____

_____

_____

Real-World Link

Lizards Two of the most common lizards people keep as pets are the chameleon and the green iguana. The chameleon feeds mostly on insects, while the green iguana is an herbivore, or feeds only on plants. It costs about $8 per month to buy food for a pet lizard at a local pet store. An online store charges an initial fee of $10, and then $6 per month for lizard food.

1. Write an equation to represent the total cost y of buying lizard food at a local pet store for any number of months x. _____

2. Write an equation to represent the total cost y of buying lizard food online for any number of months x. _____

3. Write an expression to find the cost to buy lizard food at the local pet store and online for 3, 4, 5, 6, and 7 months. Then find each cost.

Months	Cost at Pet Store ($)	Cost Online ($)
3		
4		
5		
6		
7		

4. At what month are the costs the same?

   _____

Notes

Solve Systems of Equations by Graphing

1. Fill in the blanks to explain how to solve a system of linear equations by graphing. Then solve the following system by graphing.

$$y = 2x$$
$$y = -x + 3$$

Step 1 Graph each line on _____

Step 2 Find the coordinates of the point where _____

Step 3 To check, replace the point coordinates into _____

Number of Solutions

Match each system description with its number of solutions.

2. different slopes

3. same slope, different y-intercepts

4. same slope, same y-intercept

a. infinitely many

b. exactly one

c. no solution

5. Write an example of a system of two linear equations with no solutions.

_____

Summary

Write 2–3 sentences to summarize the lesson.

_____

_____

_____

_____

_____

_____

_____

Rate Yourself!

How confident are you about solving systems of equations by graphing? Check the box that applies.

☐ ☐ ☐ ☐ ☐

For more help, go online to access a Personal Tutor.

Solve Systems of Equations Algebraically

Getting Started

Scan Lesson 9-7 in your textbook. List two headings you would use to make an outline of the lesson.

- _____
  _____

- _____
  _____

Real-World Link

Volleyball Alisha scored a total of 12 points in two volleyball games. In the second game, she scored 2 times as many points as the first game.

1. Let x represent the number of points that Alisha scored in the first game. Let y represent the number of points that she scored in the second game. Then write an equation to represent the total number of points that she scored in the two games. _____

2. Circle the expression that represents the number of points y that Alisha scored in the second game.

   $2x$ $2y$ $12x$

3. Use the equation from Exercise 1 and the expression from Exercise 2 to write a new equation using only x to represent the total number of points scored in the two games. _____

4. How many points did Alisha score in each game?

   Game 1: ☐ points

   Game 2: ☐ points

5. Does your answer make sense? Explain your reasoning.

   _____
   _____
   _____

Notes

Solve Systems of Equations Algebraically

Fill in the steps below to solve the system of equations by substitution.

$$y = 5x - 2$$
$$y = 4x$$

Step 1 Write the first equation.	
Step 2 Substitute 4x for y.	
Step 3 Solve for x.	
Step 4 Substitute your solution for x into one of the equations to solve for y.	
Step 5 Write the solution as an ordered pair. Check it in both equations.	

Interpret Solutions

There are 14 birds at a birdbath. There are 4 more bluebirds than cardinals. Suppose a system of equations was used to represent and solve the problem. Describe what the solution (x, y) would represent.

_____

_____

Summary

Write 2–3 sentences to summarize the lesson.

_____

_____

_____

_____

_____

_____

Rate Yourself!

☐ I understand how to solve a system of equations algebraically.

▶▶ Great! You're ready to move on!

☐ I still have questions about how to solve a system of equations algebraically.

▯ No Problem! Go online to access a Personal Tutor.

Chapter Review

Vocabulary Check

Complete the crossword puzzle using the vocabulary list at the beginning of the chapter.

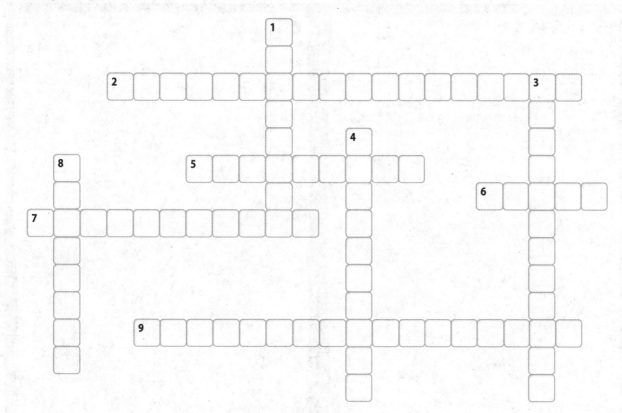

Across

2. an equation written in the form $y = mx + b$

5. In a direct variation equation, the m is the constant of _____.

6. the ratio of the vertical change to the horizontal change of a line

7. In a linear equation, the _____ variable is a variable for the input.

9. a set of equations with the same variables

Down

1. a relation in which each value of the domain is paired with exactly one value in the range

3. a description of how one quantity changes in relation to another quantity

4. what the variable b represents in the equation $y = mx + b$

8. The rate of change is _____ when the rate of change between any two data points is the same.

Use Your FOLDABLES

Use your Foldable to help review the chapter.

Tape here

Linear Functions

Example

Picture

Example

Picture

Got it?

Circle the correct choice.

1. The number of baked goods sold at a bake sale determines the amount of money earned. The money earned is the (dependent, independent) variable.

2. Slope is the ratio of (the rise to the run, the run to the rise).

3. Direct variations (are, are not) proportional relationships.

4. In the equation $y = 4x + 5$, the y-intercept is (4, 5).

5. Intersecting lines have (0, 1, infinitely many) points of intersection.

Problem Solving

1. Maria ran 17 miles at an average speed of 5 miles per hour. (Lesson 9-1)

   a. Use function notation to write an equation that gives the total distance she ran as a function of the total time. _____

   b. Use the function to find the total time she ran. _____

2. Dontae and his friends are renting m movies and buying b boxes of popcorn. Find two solutions of $5m + 1.25b = 20$. Explain each solution. (Lesson 9-2)

   _____

   _____

3. The amount of water used in an area varies directly with the population. About 18 million people in Florida use 2.4 trillion gallons of water a year. (Lesson 9-4)

   a. Write a direct variation equation relating the population x and the amount of water used y. _____

   b. Estimate the amount of water that will be needed for 24 million people.

   _____

4. A gym charges a $59 registration fee plus $7.70 per week that a person attends. The total cost y for any number of weeks x can be given by the equation $y = 7.7x + 59$. (Lesson 9-5)

   a. State the slope and y-intercept. Then graph the equation using the slope and y-intercept. _____

   b. Describe what the slope and y-intercept represent.

   _____

   _____

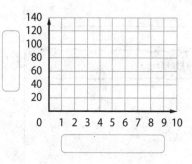

5. Rachael spent $8 for 1 magazine and 2 puzzle books. At the same store, Christopher spent $23 for 4 magazines and 4 puzzle books. (Lesson 9-7)

   a. Write a system of equations to represent this situation.

   _____

   b. Solve the system of equations algebraically. Explain what the solution means. _____

   _____

Reflect

Answering the Essential Question

Use what you learned about linear functions to complete the graphic organizer. List three ways in which functions are used to model proportional relationships. Then give an example of each.

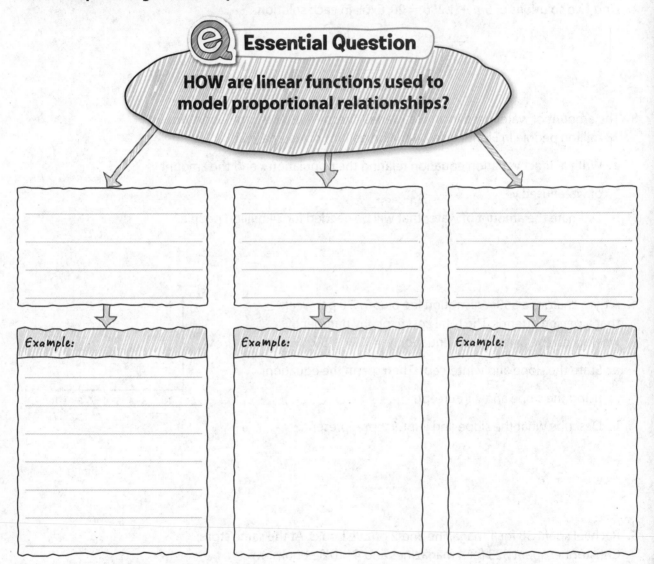

Essential Question

HOW are linear functions used to model proportional relationships?

Example:

Example:

Example:

Answer the Essential Question HOW are linear functions used to model proportional relationships?

Statistics and Probability

Chapter Preview

Vocabulary

biased sample

box plot

complement

compound event

convenience sample

distribution

double box plot

experimental probability

first quartile

Fundamental Counting Principle

interquartile range

mean absolute deviation

measures of center

measures of variability

outcome

outliers

population

probability

quartiles

random

range

relative frequency

sample

sample space

simple event

simple random sample

simulation

statistics

stratified random sample

systematic random sample

theoretical probability

third quartile

tree diagram

unbiased sample

uniform probability model

visual overlap

voluntary response sample

Vocabulary Activity

Complete the graphic organizer.

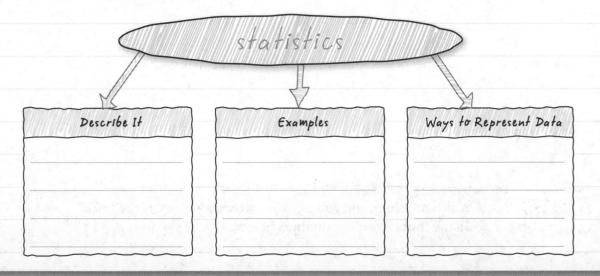

statistics

Describe It	Examples	Ways to Represent Data

Try the Quick Check below.
Or, take the Online Readiness Quiz.

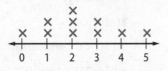

Common Core Review 6.SP.4, 6.SP.5

Example 1

Ten students were surveyed about the number of books they read over the summer. The students read 1, 5, 0, 2, 2, 1, 3, 4, 2, and 3 books. Make a line plot of the data.

Draw and label a number line. Place as many Xs above each number as there are responses for that number. Include a title.

Books Read Over Summer

```
              ×
        ×     ×
    ×   ×     ×     ×
×   ×   ×  ×  ×  ×  ×
+---+---+---+---+---+---+
0   1   2   3   4   5
```

Example 2

The high temperatures for one week were 65°, 70°, 75°, 70°, and 65°. Find the mean high temperature.

$$\frac{65 + 70 + 75 + 70 + 65}{5} = \frac{345}{5}$$ Add data. Divide number of data.

$$= 69$$ Simplify.

The mean high temperature was 69°.

Quick Check

The table shows the number of students in one class who drank chocolate milk at lunch each day for three weeks.

1. **Line Plot** Make a line plot of the data.

   Chocolate Milk

   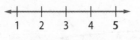

   ```
   +---+---+---+---+
   1   2   3   4   5
   ```

Chocolate Milk				
1	5	2	3	4
2	3	1	2	5
5	2	5	4	1

2. **Mean** Find the mean number of students who drank chocolate milk each day.

_____

Show your work. →

How Did You Do?

Which problems did you answer correctly in the Quick Check?
Shade those exercise numbers below.

① ②

✂ cut on all dashed lines   ▢ fold on all solid lines   tape to page 238   FOLDABLES

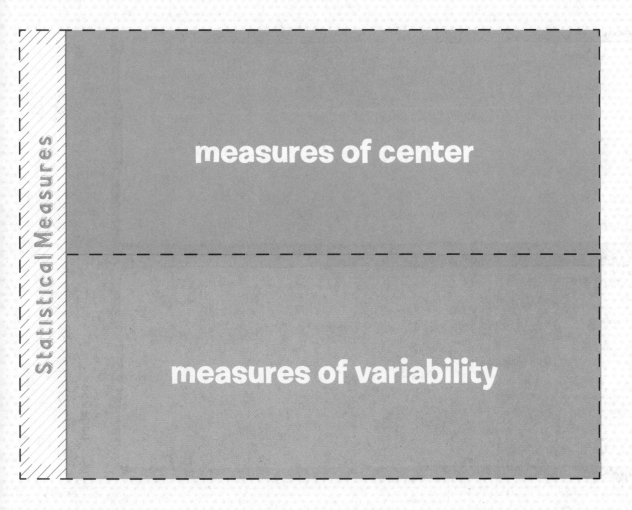

Statistical Measures

measures of center

measures of variability

FOLDABLES
Study Organizer

 Cut out the Foldable above.

 Place your Foldable on page 238.

 Use the Foldable throughout this chapter to help you learn about statistics.

Chapter 10 Foldable   **217**

Examples

Examples

page 238

Measures of Center

Getting Started

Scan Lesson 10-1 in your textbook. Write the definitions of mean and median.

• mean _____

_____

• median _____

_____

Real-World Link

Softball On average, softball players in warm weather states have faster pitching speeds than players in colder weather regions. Carly is a pitcher on her seventh grade softball team. The speeds of her last 11 pitches are shown below.

Carly's Pitching Speed (mph)					
41	39	44	38	41	38
42	39	39	40	39	

1. How would you find her average pitching speed?

_____

2. What is her average pitching speed?

       total   number of pitches   average speed

    [ ] ÷ [ ] = [ ]

3. List her pitching speeds in order from least to greatest.

_____

4. What is the middle number in the ordered data?

[ ] mph

5. Which number occurs the most often?

[ ] mph

Notes

Measures of Center

Complete the graphic organizer. Write the definition of each measure. Then show an example of how to find each value.

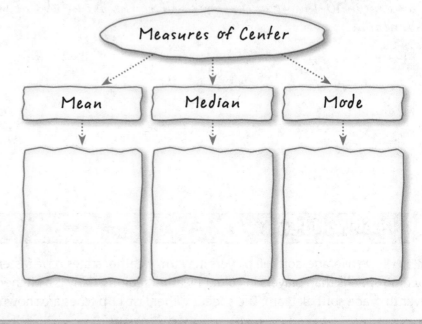

Choose Appropriate Measures

On his last nine tests, Morgan earned scores of 86, 92, 84, 78, 72, 83, 92, 90, and 79. What are the mean, median, and mode of his scores? Which measure(s) best describe his scores? Explain.

mean: ☐   median: ☐   mode: ☐

Summary

Write 2–3 sentences to summarize the lesson.

Rate Yourself!

How confident are you about finding measures of center? Check the box that applies.

For more help, go online to access a Personal Tutor.

FOLDABLES *Time to update your Foldable!*

Measures of Variability

Getting Started

Scan Lesson 10-2 in your textbook. Write the math and real-world definitions of median.

- math definition _____

  _____

- real-world defintion _____

Vocabulary Start-Up

The values that divide a set of data into four equal parts are **quartiles**. The **first quartile** is the median of the lower half of the data. The **third quartile** is the median of the upper half of the data.

Label the line plot below with the terms *median*, *lower quartile*, and *upper quartile*.

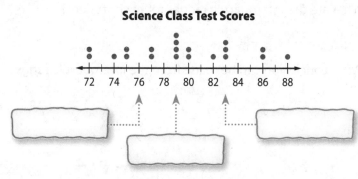

Science Class Test Scores

72 74 76 78 80 82 84 86 88

 ## Real-World Link

Spiders Spiders come in many different sizes, shapes, and colors. A spider's length is measured from its head to the end of its abdomen. The table shows lengths in millimeters of various adult spiders.

Spider Lengths (mm)		
white-tail: 15	funnel-web: 28	mouse spider: 32
black house: 15	wolf: 35	tarantula: 100

What are the mean, median, and mode of the data shown in the table?

mean: _____   median: _____   mode: _____

Notes

Measures of Variability

1. When are measures of variation used?

_____

2. What are two measures of variation? _____

Use Measures of Variability

The tables shows Jackson's and Terry's test scores.

Jackson	67	80	78	75	80	79	77	79	55
Terry	68	77	60	77	71	72	52	63	59

3. What is the range of each student's scores? _____

4. What conclusion can be made from the ranges?

_____

_____

5. What are the interquartile ranges for each student?

_____

_____

6. What conclusion can be made from the interquartile ranges?

_____

_____

Summary

Write 2–3 sentences to summarize the lesson.

_____

_____

_____

_____

_____

_____

_____

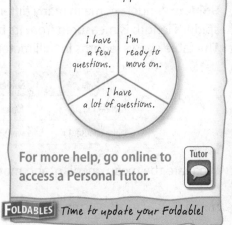

Rate Yourself!

Are you ready to move on? Shade the section that applies.

I have a few questions.

I'm ready to move on.

I have a lot of questions.

For more help, go online to access a Personal Tutor.

FOLDABLES Time to update your Foldable!

Mean Absolute Deviation

Getting Started

Scan Lesson 10-3 in your textbook. Predict two things you will learn about mean absolute deviation.

- _____

- _____

Quick Review

Evaluate the following expressions.

$|15 - 3| =$ _____

$|3 - 15| =$ _____

Real-World Link

Mammals Giraffes are the tallest land animals. Baby giraffes are between 5 and 6 feet tall, adult males are typically 18 feet tall, while adult females are around 16 feet tall. The tables show the heights of the giraffes in two different herds at a nature preserve.

Heights of Giraffes in Herd A (ft)					
6	13	8	19	9	7.5
12	6.5	14	15.5	14.5	8.5

Heights of Giraffes in Herd B (ft)						
12	13	5	5.5	7	13.5	15
16.5	14	18	11	9	6.5	10

1. Plot each set of data on the number lines below.

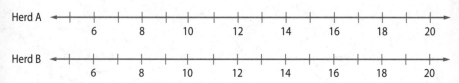

2. Find the mean of each set of data to the nearest tenth.

_____

3. Find the range of each set of data.

_____

4. Refer to the number lines and your answers to Exercises 2 and 3. Compare and contrast the sets of data.

_____

_____

_____

Questions

Notes

Find Mean Absolute Deviation

The long distance runners on the track team ran 20, 15, 25, 10, 28, and 22 total miles during practice last week. Complete the graphic organizer to find the mean absolute deviation of the miles the long distance runners ran.

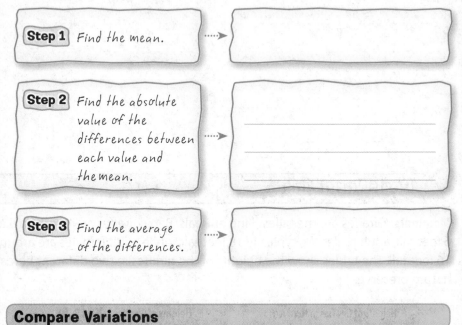

Step 1 Find the mean. ·····➔

Step 2 Find the absolute value of the differences between each value and the mean. ·····➔

Step 3 Find the average of the differences. ·····➔

Compare Variations

Complete the following statement.
A data set with a smaller mean absolute deviation has data values

that are _____ .

Summary

Write 2–3 sentences to summarize the lesson.

Rate Yourself!

How well do you understand mean absolute deviation? Circle the image that applies.

Clear   Somewhat   Not So
            Clear   Clear

For more help, go online to access a Personal Tutor.

 Tutor

Lesson 10-4
Compare Populations

Getting Started

Scan Lesson 10-4 in your textbook. List two headings you would use to make an outline of the lesson.

• _____
  _____

• _____
  _____

Quick Review

Find the mean, median, and mode of the test scores below.

81, 82, 82, 83, 83, 84, 84, 84, 85, 85, 86, 86, 87

Real-World Link

Fundraising Every year, scouts sell gourmet popcorn to raise money for troop events, including camping trips. The dot plot below shows the sales by each member of a local troop.

Individual Popcorn Sales

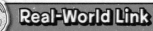
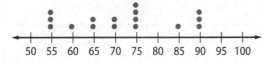

50 55 60 65 70 75 80 85 90 95 100

1. Find each of the following values.

Minimum: [ ]   Maximum: [ ]   Range: [ ]

First Quartile: [ ]   Third Quartile: [ ]   IQR: [ ]

Median: [ ]

2. What conclusions can you make from the dot plot? _____

_____

_____

3. Suppose another troop had the sales shown below.

Individual Popcorn Sales

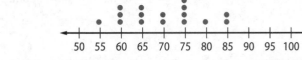

50 55 60 65 70 75 80 85 90 95 100

How does this troop's sales compare to the troop above?

_____

_____

Notes

Compare Two Populations

Use the terms *mean, median, mean absolute deviation*, and *interquartile range* to complete the table to show the most appropriate measure to use when comparing data sets. Terms may be used more than once.

	Most Appropriate Measures		
	Both sets of data are symmetric.	Neither set of data is symmetric.	Only one set of data is symmetric.
Measure of Center			
Measure of Variation			

Only One Symmetric Population

The double dot plot shows the daily low temperatures for two cities.

Daily Low Temperatures (°F)

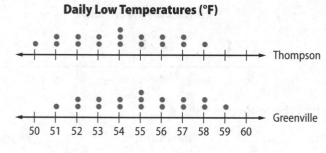

1. Is either plot symmetric? _____

2. Which measures of center and variation should you use to compare

   the data? _____

Summary

Write 2–3 sentences to summarize the lesson.

_____

_____

_____

_____

_____

_____

Rate Yourself!

Are you ready to move on? Shade the section that applies.

YES   ?   NO

For more help, go online to access a Personal Tutor.

Tutor

Mid-Chapter Check

Vocabulary Check

1. **CCSS** **Be Precise** Define *mean*. What two other measures are used to represent a whole set of data? (Lesson 1) _____

_____

Skills Check and Problem Solving

A theater showed a movie for two weeks. The table shows the number of people who attended the movie during Week 1. (Lessons 1–3)

2. What are the measures of center of the data in the table? Round to the nearest tenth, if necessary.

_____

3. What are the measures of variability of the data in Exercise 2?

_____

4. What is the mean absolute deviation of the data? Round to the nearest tenth.

_____

Week 1 Attendance	
Day	People
1	49
2	35
3	62
4	41
5	145
6	163
7	150

The table at the right shows the number of people who attended the same movie during Week 2. (Lesson 3)

5. What is the mean absolute deviation of the data? Round to the nearest tenth.

_____

6. What conclusion can you make based on the mean absolute deviations?

_____

_____

_____

Week 2 Attendance	
Day	People
1	64
2	75
3	82
4	98
5	115
6	124
7	136

7. **Standardized Test Practice** Refer to the information above about movie attendance. Which statement is an accurate comparison of the Week 1 and Week 2 movie populations? (Lesson 4)

Ⓐ Each set of data has a mode.

Ⓑ The interquartile range of the Week 1 data is less than the interquartile range of the Week 2 data.

Ⓒ Week 1 data are more spread out than Week 2 data.

Ⓓ Week 2 data are more spread out than Week 1 data.

21ST CENTURY CAREER
in Market Research

Keeping Your Eye on the Target Market!

The table shows the results of a survey about social networking site usage.

1. Find the measures of center of the data.

_____

2. Which measure of center best represents the data? Explain your reasoning.

_____

_____

3. Find the measures of variability and any outliers for the data.

_____

_____

4. Use the measures of variability to describe the data.

_____

_____

_____

5. Find the mean absolute deviation of the data set. Then describe what the mean absolute deviation represents. Round to the nearest tenth.

_____

_____

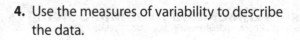

Daily Time Spent on Social Networking Sites (min)					
105	60	30	165	90	120
60	90	45	90	180	60
150	75	60	75	135	120
30	120	45	90	60	165

Career Project

It's time to update your career portfolio! Use the Internet or another source to research a career as a market research analyst. Write a paragraph that summarizes your findings.

_____

_____

_____

_____

What skills would you need to improve to succeed in this career?

- _____
- _____
- _____
- _____
- _____

Lesson 10-5
Using Sampling to Predict

Getting Started

Scan Lesson 10-5 in your textbook. Predict two things you will learn about sampling.

- _____
  _____

- _____
  _____

Vocabulary

Write the definition of *survey* in your own words.

_____

_____

_____

Vocabulary Start-Up

The group being studied in a survey is the **population**. Sometimes a population is too large to survey everyone. Then a part of the population, or **sample**, is surveyed. An **unbiased sample** is a sample that is selected at random and is representative of the larger population.

A principal wants a survey to represent all the students in the school. For each sample, write B for *biased* or U for *unbiased*.

1. Every 10th person is chosen from an alphabetical list of all students in the school. _____

2. Every student in science class is surveyed. _____

3. Students who wish to respond to the survey fill out a form and turn it in during homeroom. _____

4. Every student in the school is given a different number. Twenty numbers are chosen at random. _____

Real-World Link

School Clubs Many schools offer a variety of clubs for students. Suppose your school wanted to start a new club.

5. What are two ways you could survey students about which club they would like to see started? _____

_____

6. Do you think the ways you listed in Exercise 5 are random? Do you think they are representative of the larger population? Explain.

_____

_____

Notes

Identify Sampling Techniques

1. Cross out the part of the concept circle that does not belong.

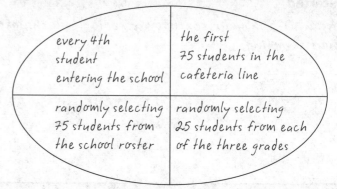

every 4th student entering the school | the first 75 students in the cafeteria line

randomly selecting 75 students from the school roster | randomly selecting 25 students from each of the three grades

2. Explain your reasoning in Exercise 1.

_____

_____

_____

Validating and Predicting Samples

3. A manufacturer makes 1500 phones and tests every 10th phone for defects. Of the phones, 24 were defective. Is the sampling valid? If yes, how many of the 1500 phones could you expect to be defective? If no, how could the manufacturer better test for defective phones?

_____

Summary

Write 2–3 sentences to summarize the lesson.

_____

_____

_____

_____

_____

_____

Rate Yourself!

How confident are you about using sampling? Shade the ring on the target.

I'm on target.

I need help.

For more help, go online to access a Personal Tutor.

Probability of Simple Events

Getting Started

Scan Lesson 10-6 in your textbook. List two real-world scenarios in which you would find probability.

- _____
  _____
- _____
  _____

Vocabulary Start-Up

Probability is the chance that an event will occur. An **outcome** is the possible result of a probability experiment.

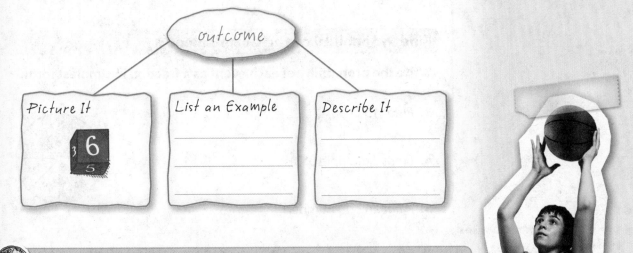

outcome

| Picture It | List an Example | Describe It |

Real-World Link

Basketball A free throw is an unguarded shot awarded when an opponent commits a foul or rule infraction. In a basketball game, Alex makes 12 out of 15 free throws attempted.

1. Write Alex's free throw percentage as a fraction in simplest form.

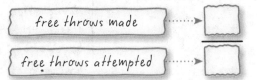

free throws made ·······▸ ☐
                          ——
free throws attempted ·······▸ ☐

2. Do you expect Alex to make the next free throw? Explain.

_____

Notes

Probability

1. Fill in each blank with the terms *certain, impossible,* or *equally likely.*

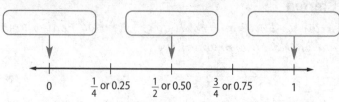

0        $\frac{1}{4}$ or 0.25 $\frac{1}{2}$ or 0.50 $\frac{3}{4}$ or 0.75   1

2. A number cube is rolled. Write an example of each likelihood.

Event	Likelihood
	equally likely
	certain
	impossible

Find Probability of the Complement

Write the probability of each event as a fraction in simplest form.

3. *P*(*not* 15 or 16) _____

4. *P*(*not* a multiple of 3) _____

5. *P*(*not* an odd number) _____

Summary

Write 2–3 sentences to summarize the lesson.

_____

_____

_____

_____

_____

_____

Rate Yourself!

How confident are you about finding probability? Check the box that applies.

For more help, go online to access a Personal Tutor.

Theoretical and Experimental Probability

Getting Started

Scan Lesson 10-7 in your textbook. List two headings you would use to make an outline of the lesson.

- _____

_____

- _____

_____

Vocabulary
Write the definition of *probability* in your own words.

_____

_____

_____

_____

Real-World Link

Games Two different versions of Ticket Extravaganza are shown below. The object is to slide a token through a shoot and have it land on one of the tiles to win the amount of tickets shown.

Game A

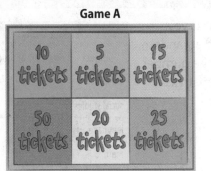

Game B

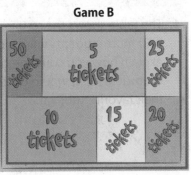

1. Are you more likely to land on 50 tickets in Game A or Game B? Explain.

_____

_____

2. In a uniform probability model, each outcome has an equal probability of happening. Which game has uniform probability? _____

3. Copy each game board onto a separate piece of paper. Then cut out a 1-inch square piece of scrap paper and roll it into a ball. Drop the ball onto each game board 5 times. Record your results.

4. In which game would you earn more tickets? Explain.

_____

Turn	Game A	Game B
1		
2		
3		
4		
5		

Notes

Make Predictions

The graph shows the results of an experiment in which a spinner with 3 equal sections is spun fifty times.

1. What is the theoretical probability of

   spinning green? _____

2. What is the experimental probability of

   spinning green? _____

3. What is the difference between theoretical
   probability and experimental probability?

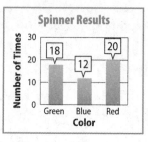

Spinner Results

_____

_____

_____

Make Predictions

4. The table shows the results of a survey.
   If 50 more students were picked at random,
   explain how you could predict the number
   of students that would choose math as
   their favorite subject.

Favorite Subject	
Subject	**Number of Students**
science	22
social studies	22
language arts	26
math	30

_____

_____

_____

_____

Summary

Write 2–3 sentences to summarize the lesson.

_____

_____

_____

_____

_____

_____

Rate Yourself!

How well do you understand theoretical and experimental probability? Circle the image that applies.

Clear   Somewhat   Not So
              Clear   Clear

For more help, go online to access a Personal Tutor.

Probability of Compound Events

Getting Started

Scan Lesson 10-8 in your textbook. Predict two things you will learn about finding the probability of compound events.

- _____
  _____

- _____
  _____

Real-World Link

Pizza Pizza is a favorite food in many American households because you can get just about anything on a pizza. A certain pizzeria offers a create-your-own pizza special with the options shown in the table.

Size	Crust	Toppings (choose one)
10 in. 12 in. 14 in.	thick thin	pepperoni sausage onions mushrooms

1. Make a list of all the 10-inch pizzas that can be created.

   How many different pizzas 10-inch pizzas can be created? _____

   If a 10-inch pizza is ordered randomly, what is the probability of ordering it with a thin crust, and pepperoni? _____

2. How many different pizzas of any size can be created? _____

3. If a type of pizza is ordered randomly, what is the probability of ordering a 10-inch, thin crust, with pepperoni? _____

   What is the probability of ordering any size pizza with thick crust topped with a vegetable? _____

4. In Exercise 1, you made a list to show all of the different 10-inch pizzas. Suppose you wanted to find all of the different pizzas. What is another way to show all of the pizzas? _____

Notes

Outcomes of Compound Events

An ice cream shop has the following options for sundaes. Describe a method you could use to find all of the possible outcomes for selecting one flavor of ice cream, one sauce, and one topping. Then use that method to find the total number of outcomes.

Flavor	Sauce	Toppings
vanilla	fudge	whipped cream
chocolate	pineapple	fruit
strawberry	caramel	

Probability of Compound Events

Find each probability using a number cube labeled 1 through 6.

1. What is the probability of tossing a 1 and then a 2? _____

2. What is the probability of tossing a 4 on two consecutive tosses?

_____

Summary

Write 2–3 sentences to summarize the lesson.

Rate Yourself!

☐ *I understand how to find the probability of compound events.*

▶▶ Great! You're ready to move on!

☐ *I still have questions about finding the probability of compound events.*

📖 No Problem! Go online to access a Personal Tutor. Tutor

Chapter Review

Vocabulary Check

Unscramble each of the clue words. After unscrambling all of the terms, use the numbered letters to find the phrase.

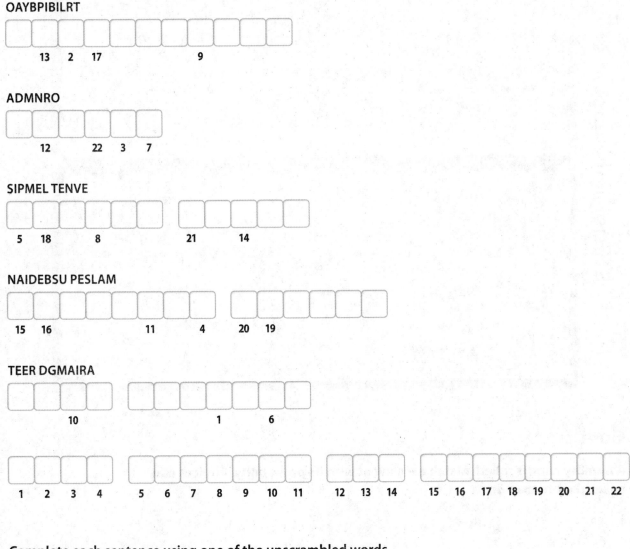

OAYBPIBILRT

☐☐☐☐☐☐☐☐☐☐☐
13 2 17   9

ADMNRO

☐☐☐☐☐☐
12 22 3 7

SIPMEL TENVE

☐☐☐☐☐☐   ☐☐☐☐☐
5 18 8   21   14

NAIDEBSU PESLAM

☐☐☐☐☐☐☐☐   ☐☐☐☐☐☐
15 16   11 4   20 19

TEER DGMAIRA

☐☐☐☐☐   ☐☐☐☐☐☐
  10   1   6

☐☐☐☐ ☐☐☐☐☐☐☐ ☐☐☐ ☐☐☐☐☐☐☐☐
1 2 3 4 5 6 7 8 9 10 11 12 13 14 15 16 17 18 19 20 21 22

Complete each sentence using one of the unscrambled words.

1. A(n) _____ is one outcome or a collection of outcomes.

2. Outcomes occur at _____ if each outcome is equally likely to occur.

3. A(n) _____ is a random sample that is representative of a larger sample.

4. _____ is the chance that an event will occur.

Use Your FOLDABLES

Use your Foldable to help review the chapter.

Tape here

Statistical Measures

Explanation

Explanation

Got it?

A number cube is rolled. Match each event with its probability. Choices may be used more than once.

1. $P(3)$ **a.** $\frac{1}{6}$

2. $P(\text{even number})$ **b.** $\frac{1}{3}$

3. $P(2 \text{ or } 3)$ **c.** $\frac{1}{2}$

4. $P(\text{prime number})$ **d.** $\frac{2}{3}$

5. $P(\text{number starting with the letter } t)$ **e.** $\frac{5}{6}$

6. $P(\text{multiple of } 3)$ **f.** 1

Problem Solving

1. Mrs. Thomas teaches two pre-algebra classes. The tables show the most recent test scores for both classes.

First Period							
92	98	96	84	90	82	64	76
72	90	90	98	74	72	86	88

Second Period							
90	98	76	74	92	80	82	84
70	60	80	92	94	78	80	90

   a. Find the mean and mode of the data for each class. (Lesson 1)

   b. Find the measures of variation for the two classes. (Lesson 2)

2. The Walnut Springs Middle School cafeteria staff wants to know which types of pizza to serve. The staff surveys every 5th student that walks into the cafeteria about their pizza preference. Identify the sample as biased or unbiased and

   describe its type. (Lesson 5) _____

3. Of the 28 socks in Maliyah's sock drawer, 8 are black, 6 are brown, 4 are gray, and the rest are white. What is the probability she will pull a white sock out of the drawer without looking? Write in simplest form. Explain your reasoning.
   (Lesson 6)

4. Mr. Gomez surveyed his class to see which sports they preferred playing. Forty-one percent preferred football, 24% basketball, 21% volleyball, and 14% track and field. Out of 620 students in the entire school, how many would you

   expect to say they preferred playing basketball? (Lesson 7) _____

5. The Willis family is planning a vacation. A travel agency offers a vacation package which allows you to choose one hotel, one car, and two activities from the list at the right. How many different combinations of hotel, car, and two activities can

   be made? (Lesson 8) _____

Hotels	Cars	Activities
Sandy Beach	SUV	Parasailing
Ocean Waves	Convertible	Fishing
The Dunes		Surfboarding
		Snorkeling

Reflect

 Answering the Essential Question

Use what you learned about statistics to complete the graphic organizer.
Describe how statistics are used to draw inferences and compare populations.

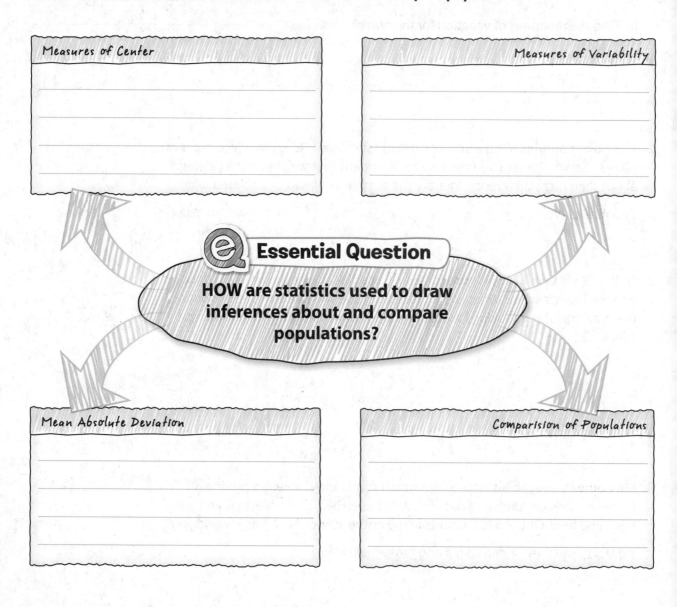

Measures of Center

Measures of Variability

 Essential Question

HOW are statistics used to draw inferences about and compare populations?

Mean Absolute Deviation

Comparision of Populations

 Answer the Essential Question. HOW are statistics used to draw inferences about and compare populations?

Chapter 11

Congruence, Similarity, and Transformations

Chapter Preview

Vocabulary

adjacent angles	image	rotational symmetry
alternate exterior angles	interior angle	supplementary angles
alternate interior angles	line of reflection	tessellation
center of rotation	line segment	transformation
complementary angles	parallel lines	translation
congruent	perpendicular lines	transversal
corresponding angles	polygon	triangle
diagonal	reflection	vertex
dilation	regular polygon	vertical angles
exterior angle	rotation	

Vocabulary Activity

Select four vocabulary terms from the list above. Write a definition for the terms based upon what you already know. As you go through the chapter, come back to this page and update your definitions if necessary.

Term	What I Know	What I Learned

Try the Quick Check below.
Or, take the Online Readiness Quiz.

Check ✓

CCSS Quick Review

Common Core Review 6.G.3, 7.EE.4

Example 1

Rectangle *ABCD* has vertices $A(-3, 2)$, $B(5, 2)$, $C(5, -4)$, and $D(-3, -4)$. Graph the rectangle.

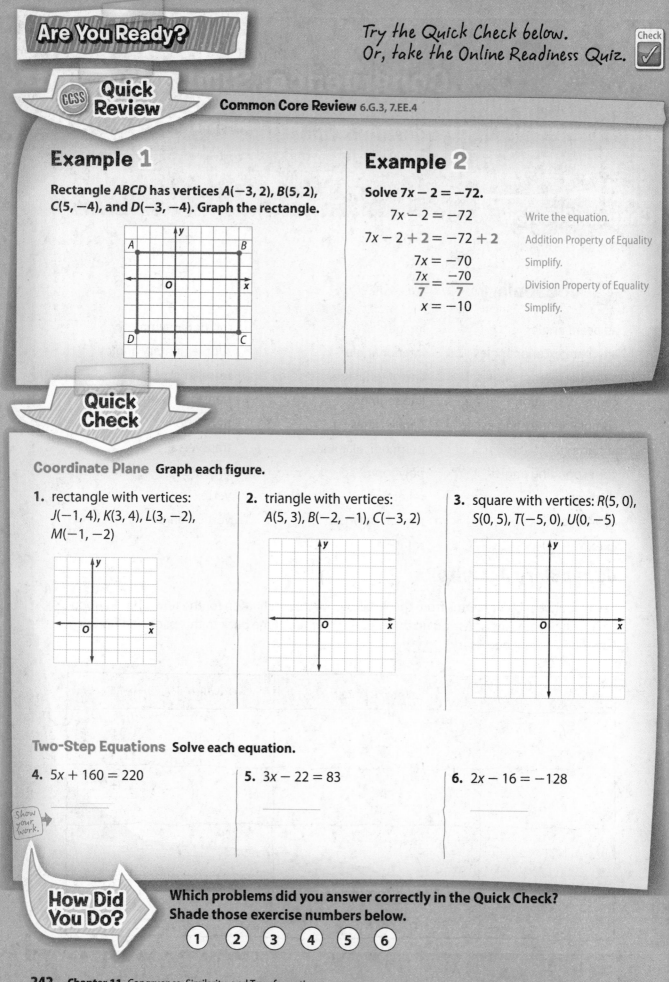

Example 2

Solve $7x - 2 = -72$.

$7x - 2 = -72$	Write the equation.
$7x - 2 + 2 = -72 + 2$	Addition Property of Equality
$7x = -70$	Simplify.
$\dfrac{7x}{7} = \dfrac{-70}{7}$	Division Property of Equality
$x = -10$	Simplify.

Quick Check

Coordinate Plane Graph each figure.

1. rectangle with vertices:
$J(-1, 4)$, $K(3, 4)$, $L(3, -2)$, $M(-1, -2)$

2. triangle with vertices:
$A(5, 3)$, $B(-2, -1)$, $C(-3, 2)$

3. square with vertices: $R(5, 0)$, $S(0, 5)$, $T(-5, 0)$, $U(0, -5)$

Two-Step Equations Solve each equation.

4. $5x + 160 = 220$

5. $3x - 22 = 83$

6. $2x - 16 = -128$

Show your work.

How Did You Do?

Which problems did you answer correctly in the Quick Check?
Shade those exercise numbers below.

① ② ③ ④ ⑤ ⑥

cut on all dashed lines   fold on all solid lines   tape to page 264   **FOLDABLES**

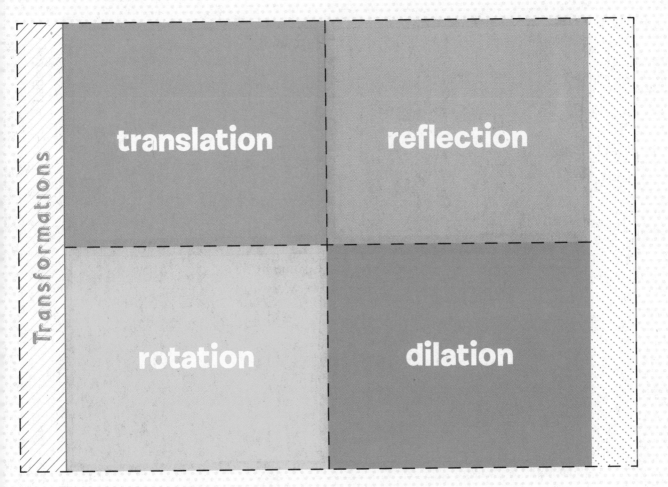

Transformations

translation

reflection

rotation

dilation

FOLDABLES
Study Organizer

1 Cut out the Foldable above.

2 Place your Foldable on page 264.

3 Use the Foldable throughout this chapter to help you learn about transformations.

page 264

Symbols

Symbols

page 264

Symbols

Symbols

Tab 2

Tab 1

Angle and Line Relationships

Getting Started

Scan Lesson 11-1 in your textbook. List two headings you would use to make an outline of the lesson.

- _____
  _____
- _____

Vocabulary

Circle the vocabulary word defined below.

Two rays that meet at a common endpoint.

angle   line

Vocabulary Start-Up

Vertical angles are pairs of opposite angles formed by intersecting lines.
Adjacent angles are two angles that have the same vertex, share a common side, and do not overlap.

Complementary angles are two angles with measures that add up to 90°.
Supplementary angles are two angles with measures that add up to 180°.

Draw an example of each type of angle pair in the organizer below.

Vertical

Adjacent

angles

Complementary

Supplementary

🌎 Real-World Link

Roller Coasters Many steel roller coasters have a 90-degree angle of descent. Several coasters in Europe boast descents of 97 degrees, while two coasters in the United States have descents of 95 degrees.

Identify the type of angle. Write *acute*, *right*, or *obtuse*.

1. 90° _____

2. 97° _____

Notes

Pairs of Angles

Determine whether each is *true* or *false*. If *false*, correct the sentence.

1. The sum of the measures of complementary angles is 180°.

   _____

   _____

2. Vertical angles are adjacent.

   _____

3. The intersection of perpendicular lines forms a right angle. _____

Parallel Lines

In the figure below, line *a* is parallel to line *b*.

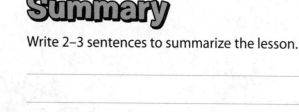

4. Draw a transversal that intersects lines *a* and *b*. Then number the angles formed by the intersection of the transversal and the lines.

5. Name a pairs of congruent angles in the figure. Then describe and classify the relationship between the angles.

   _____

   _____

Summary

Write 2–3 sentences to summarize the lesson.

_____

_____

_____

_____

_____

_____

_____

Rate Yourself!

How confident are you about identifying angle relationships? Shade the ring on the target.

For more help, go online to access a Personal Tutor.

Triangles

Getting Started

Scan Lesson 11-2 in your textbook. Write the definitions of interior angle and exterior angle.

- interior angle _____

  _____

- exterior angle _____

  _____

Real-World Link

Architecture Triangles are used in the design of many structures, including buildings, bridges, and amusement park rides. The support structure of the truss bridge below is made of many different triangular units.

1. Copy triangle *ABC* onto a separate piece of paper three times. Cut out all three triangles. Write *A*, *B*, and *C* inside the appropriate angle of each triangle. Arrange the three triangles so that each angle meets at one point. Draw a sketch of the position of the three triangles.

2. What type of angle is formed where the three angles meet? _____

3. Use a straightedge to draw another triangle. Repeat the activity. What type of angle is formed where the three angles meet? _____

4. How do the angles for Exercises 2 and 3 compare?

   _____

   What does this tell you about the sum of the measures of the angles of a triangle? _____

Notes

Angle Sum of a Triangle

Complete the graphic organizer to determine the angle measures of triangle *KLM* with a ratio of 1:3:1.

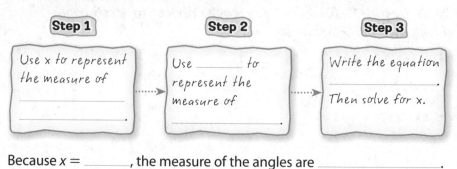

| Step 1 | Step 2 | Step 3 |

Use *x* to represent the measure of _____ _____.

Use _____ to represent the measure of _____.

Write the equation _____. Then solve for *x*.

Because *x* = _____, the measure of the angles are _____.

Classify Triangles

Draw and label each type of triangle.

Classify Triangles by Sides		
scalene	isosceles	equilateral

Classify Triangles by Angles		
acute	right	obtuse

Summary

Write 2–3 sentences to summarize the lesson.

_____

_____

_____

_____

_____

_____

Rate Yourself!

How confident are you about triangles? Check the box that applies.

☹ ☹ ☺

☐ ☐ ☐ ☐ ☐

For more help, go online to access a Personal Tutor.

Tutor

Polygons

Getting Started

Scan Lesson 11-3 in your textbook. List two headings you would use to make an outline of the lesson.

- _____
  _____

- _____
  _____

Vocabulary
Write the definition of *quadrilateral* in your own words.

Real-World Link

Polygons While riding in a car, you might see several different sizes and shapes of road signs. Some examples of typical road signs are shown below.

| Sign 1 | Sign 2 | Sign 3 | Sign 4 | Sign 5 | Sign 6 |

1. The signs can be sorted into two different groups: signs with curved edges and signs with straight edges. List the signs that fit into each group.

   Signs with Curved Edges   Signs with Straight Edges

   _____   _____

   _____   _____

   _____   _____

2. For the signs that have straight edges, how many sides does each sign have?

   _____

3. A *polygon* is a simple, closed figure formed by three or more line segments. Which group of signs is considered polygons?

   _____

4. Draw an example of a polygon and an example of a non polygon.

Notes

Classify Polygons

Mark an X over the figures that are *not* polygons. If a figure is not a polygon, write the reason below the figure.

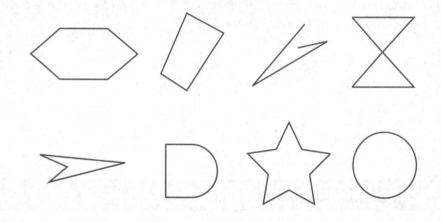

Interior Angles of a Polygon

Complete each step in the graphic organizer to find the sum of the measures of the interior angles of an octagon.

Step 1	Write the equation for a polygon with n sides.	
Step 2	Determine the number of sides in an octagon.	
Step 3	Replace n in the equation with the number of sides.	
Step 4	Simplify.	

Summary

Write 2–3 sentences to summarize the lesson.

_____

_____

_____

_____

_____

_____

Rate Yourself!

How well do you understand polygons? Circle the image that applies.

Clear   Somewhat Clear   Not So Clear

For more help, go online to access a Personal Tutor.

Translations and Reflections on the Coordinate Plane

Getting Started

Write the math and real-world definitions of transform.

- real-world definition: _____

_____

- math definition: _____

Vocabulary Start-Up

A **transformation** is a movement of a geometric figure. **Translations** and **reflections** are two types of transformations.

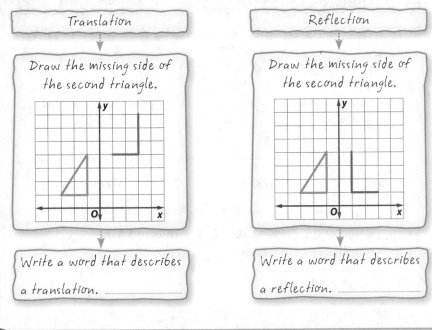

Translation

Draw the missing side of the second triangle.

Write a word that describes a translation. _____

Reflection

Draw the missing side of the second triangle.

Write a word that describes a reflection. _____

Real-World Link

Reflections A mirror image is something that appears identical, but in reverse. It can also be called a reflection. Milena is creating a reflection of the letter E. Circle the image that Milena should use.

E - - - → E   E ⋮ Ǝ

Notes

Translations

1. Write the vertices of the image of triangle *PQR* after a translation 3 units left and 1 unit down. Graph the translation.

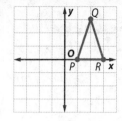

P' (_____ , _____)

Q' (_____ , _____)

R' (_____ , _____)

2. Suppose a quadrilateral is translated *a* units right and *b* units up. Explain how to find the vertices of its image.

_____

_____

Reflections

3. Fill in the blanks to complete the steps for drawing a reflection.

Step 1 Count the number of units between each _____ and the _____ .

Step 2 For each vertex, plot a _____ the same _____ away from the line on the other side.

Step 3 Connect the new _____ to form the _____ .

4. Suppose triangle *PQR* in Exercise 1 is reflected over the *y*-axis. Write the coordinates of its image.

_____

Summary

Write 2–3 sentences to summarize the lesson.

_____

_____

_____

_____

_____

_____

Rate Yourself!

Are you ready to move on? Shade the section that applies.

YES   ?   NO

For more help, go online to access a Personal Tutor.   Tutor

FOLDABLES Time to update your Foldable!

Mid-Chapter Check

Vocabulary Check

1. **CCSS** **Be Precise** Define *transformation*. Identify two types of transformations. (Lesson 4)

_____

Skills Check and Problem Solving

Find the sum of the measures of the interior angles of each polygon. (Lessons 2 and 3)

2. triangle _____

3. quadrilateral _____

4. 14-gon _____

Graph each polygon with the given vertices. Then graph the image after the given transformation and write the coordinates of the image's vertices. (Lesson 4)

5. Translate triangle *ABC* with vertices *A*(3, 5), *B*(4, 1), and *C*(1, 2) 2 units left and 5 units down.

6. Reflect rectangle *RSTU* with vertices *R*(2, 1), *S*(2, 4), *T*(4, 4), *U*(4, 1) over the *x*-axis

7. **Standardized Test Practice** In the figure, *m* ∥ ℓ and *t* is a transversal. If *m*∠3 = 140° , what are two other angles that equal 140°? (Lesson 1)

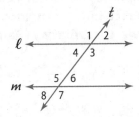

Ⓐ ∠1 and ∠2

Ⓑ ∠2 and ∠3

Ⓒ ∠5 and ∠7

Ⓓ ∠5 and ∠8

An Animation Sensation

Use Figures 1 and 2 to solve each problem.

1. In Figure 1, line *AB* is parallel to line *CD* and line *BC* is a transversal. Classify the relationship between ∠*ABC* and ∠*BCD*.

   _____

2. Classify triangle *XYZ* in Figure 1 by its angles and by its sides. _____

   _____

3. Classify polygon *JKLMN* in Figure 1. Is it a regular polygon? Explain. _____

   _____

   _____

4. In Figure 2, the car is translated 8 units left and 5 units down so that it appears to be moving. What are the coordinates of *A'* and *B'* after the

   translation? _____

5. In Figure 2, the car is translated so that *A'* has coordinates (−7, 2). Describe the translation as an ordered pair. Then find the coordinates of

   point *B'*. _____

   _____

6. In Figure 2, the car is reflected over the *x*-axis in order to make its reflection appear in a pond. What are the coordinates of *A'* and *B'*

   after the reflection? _____

   _____

Figure 1

Figure 2

Career Project

It's time to update your career portfolio! Choose a movie that was completely or partially computer animated. Use the Internet to research how technology was used to create the scenes in the movie. Describe any challenges that the computer animators faced.

_____

_____

_____

_____

_____

What are some short term goals you need to achieve to become a computer animator?

- _____
- _____
- _____
- _____
- _____

Rotations on the Coordinate Plane

Getting Started

Scan Lesson 11-5 in your textbook. Predict two things you will learn about rotations.

- _____
- _____

Vocabulary
Write the definition of rotation in your own words.

_____
_____
_____

Real-World Link

Fairs There are many things to see and do at the fair. You can look at the animals, browse arts and crafts projects, and ride the carnival rides! Some rides move up and down, while others, like the carousel, move around in a circle. You sit on a carousel horse and ride around in a circle many times.

1. Does the size and shape of the carousel change as it spins? _____

2. The carousel spins in a *counterclockwise* direction. Define the terms *clockwise* and *counterclockwise* in your own words.

clockwise _____

_____

counterclockwise _____

_____

3. The drawing shows a view of the carousel from above. Point *A* represents Kendra's position at the beginning of the ride. Point *B* shows her position after 5 seconds.

Describe the movement involved as Kendra travels from point *A* to point *B*.

_____

4. How would you classify $\angle ACB$?

_____

What is the measure of this angle? _____

Notes

Rotations

Label each diagram with the correct angle of rotation about the origin.

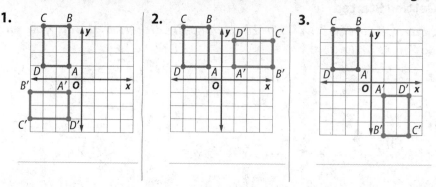

1.   **2.**   **3.**

Rotational Symmetry

Fill in the graphic organizer about rotational symmetry.

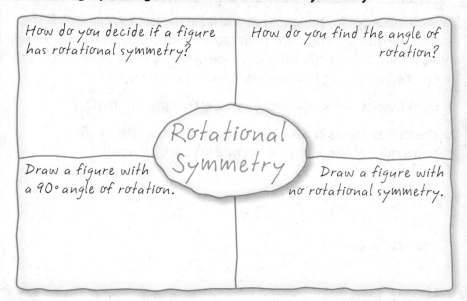

How do you decide if a figure has rotational symmetry?

How do you find the angle of rotation?

Rotational Symmetry

Draw a figure with a 90° angle of rotation.

Draw a figure with no rotational symmetry.

Summary

Write 2–3 sentences to summarize the lesson.

Rate Yourself!

Are you ready to move on? Shade the section that applies.

YES   ?   NO

For more help, go online to access a Personal Tutor.

FOLDABLES Time to update your Foldable!

Congruence and Transformations

Getting Started

Scan Lesson 11-6 in your textbook. List two real-world scenarios in which you would use transformations.

- _____
  _____

- _____
  _____

Real-World Link

Fabrics Interior designers often use fabrics with repeatable patterns. These patterns are created by transforming geometric shapes. The fabric shown below is an example. It contains a pattern that is made up of green, yellow, purple, and blue triangles.

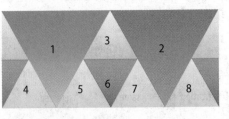

Copy the figure above on tracing paper.

1. Cut out triangle 1 and place it on top of triangle 2. What do you notice about triangles 1 and 2? _____

2. What does that tell you about triangles 1 and 2? _____

3. What transformation(s) could you use to map triangle 1 onto triangle 2?
   _____

4. Cut out triangle 3. What triangles have the same size and shape as triangle 3? _____

5. What transformation(s) could you use to map triangle 3 onto triangle 6? _____

6. Place triangle 3 on top of triangle 1. What do you notice?
   _____
   _____

Notes

Identify Congruency

1. Cross out the figure(s) that are not congruent to the figure at the right.

2. How can you use transformations to show that two figures are

   congruent? _____

   _____

Identify Transformations

3. In the table, circle the words that describe the length and orientation of a figure and its image.

Translation		
Length	same	different
Orientation	same	reversed	changed
Reflection		
Length	same	different
Orientation	same	reversed	changed
Rotation		
Length	same	different
Orientation	same	reversed	changed

Summary

Write 2–3 sentences to summarize the lesson.

_____

_____

_____

_____

_____

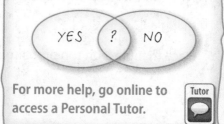

Rate Yourself!

Are you ready to move on? Shade the section that applies.

YES   ?   NO

For more help, go online to access a Personal Tutor.

Dilations on the Coordinate Plane

Getting Started

Scan Lesson 11-7 in your textbook. List two real-world scenarios in which you would use dilations.

*   _____

    _____

*   _____

    _____

Quick Review

What is the scale factor on a scale model if the scale is 1 inch = 3 feet?

_____

Real-World Link

Photography Whether you print your own photographs, or purchase them from someone else, you can get them printed in just about any size you want. Anya has a 5-inch by 7-inch photo of her family that she wants to print in different sizes to give as gifts.

1. Anya wants to have a 10-inch by 14-inch portrait printed for her parents. Is this an *enlargement* or *reduction* of the original picture? _____

   What is the scale factor from the original photo to the new photo?

   scale factor: $\dfrac{\text{length on new picture}}{\text{original length}} = \dfrac{\square}{\square}$ or $\dfrac{\square}{\square}$

2. Anya is using pictures that are 2.5-inches by 3.5-inches for keychains. Is this an *enlargement* or *reduction* of the original picture? _____

   What is the scale factor from the original photo to the new photo?

   scale factor: $\dfrac{\text{length on new picture}}{\text{original length}} = \dfrac{\square}{\square}$ or $\dfrac{\square}{\square}$

3. Identify whether each pair of dimensions indicates an enlargement or reduction. Then find the scale factor from the original photo to the new photo for each pair of dimensions.

   4 in. × 6 in. to 10 in. × 15 in. _____

   8 in. × 10 in. to 2 in. × 2.5 in. _____

4. What can you conclude about the scale factors of enlargements and reductions? _____

   _____

Notes

Dilations

A figure has vertices *J*(1, 1), *K*(3, 1), *L*(3, 2) and *M*(2, 3). Graph the figure and the image of the polygon after a dilation with a scale factor of 2.

J(1, 1) → *J'*(⬚ , ⬚)

K(3, 1) → *K'*(⬚ , ⬚)

L(3, 2) → *L'*(⬚ , ⬚)

M(2, 3) → *M'*(⬚ , ⬚)

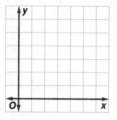

Find a Scale Factor

On the graph, one figure is a dilation of the other. Find the scale factor of the dilation and classify it as an *enlargement* or as a *reduction*.

Write a ratio of each *x*- or *y*-coordinate of one vertex of the dilation to the *x*- or *y*-coordinate of the corresponding vertex of the original figure.

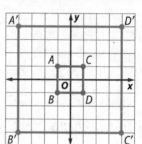

scale factor: _____

type of dilation: _____

Summary

Write 2–3 sentences to summarize the lesson.

_____

_____

_____

_____

_____

_____

Similarity and Transformations

Getting Started

Scan Lesson 11-8 in your textbook. List two headings you would use to make an outline of the lesson.

* _____

_____

* _____

_____

Quick Review

Write the definition of *similar figures* in your own words.

Real-World Link

Art Artists use grids to transfer their work to a larger area. A grid ensures that what an artist draws or paints retains the correct proportions when it is enlarged.

1. Reflect the triangle over the dashed line. Then dilate the new triangle using a scale factor of $\frac{1}{2}$. Label the corresponding sides of the final triangle a', b', and c'.

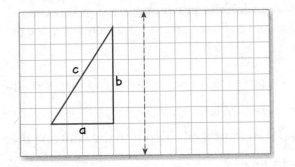

2. Fill in the boxes to write ratios comparing side lengths in simplest form.

Original Triangle

$$\frac{a}{b} = \frac{\square}{\square} = \frac{\square}{\square}$$

Final Triangle

$$\frac{a'}{b'} = \frac{\square}{\square}$$

3. Use a ruler to find c and c' to the nearest millimeter. Then find $\frac{b}{c}$ and $\frac{b'}{c'}$.

What do you notice? _____

4. Is the final triangle similar to the original triangle? Explain your reasoning.

_____

5. If a dilation is in a series of transformations, does the order of the

transformations affect the shape? Explain. _____

_____

Notes

Identify Similarity

Describe each set of transformations. Then determine whether each set of transformations produces a figure that is similar to the original. Write *yes* or *no*.

1.

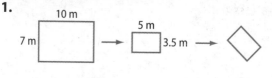

10 m
7 m
5 m
3.5 m

_____

2.

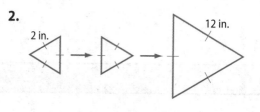

2 in.
12 in.

_____

Use the Scale Factor

Determine whether sentence produces a figure that is similar to the original figure. Write *yes* or *no*.

3. A trapezoid is dilated with a scale factor of $\frac{1}{3}$ and then rotated 180°. _____

4. Two sides of a square are doubled and then the figure is reflected across a vertical line. _____

5. A triangle is translated six units to the right and then dilated with a scale factor of 2. _____

Summary

Write 2–3 sentences to summarize the lesson.

_____

_____

_____

_____

_____

_____

_____

_____

Rate Yourself!

How confident are you about dilations? Shade the ring on the target.

I'm on target.

I need help.

For more help, go online to access a Personal Tutor.

Tutor

Chapter Review

Vocabulary Check

Fill in the blank with the correct vocabulary term. Then circle the word that completes the sentence in the word search.

1. _____ angles are two angles that have the same vertex between them, share a common side, and do not overlap.

2. A _____ is when you slide a figure from one position to another without turning it.

3. Two angles are _____ if the sum of their measures is 180°.

4. A _____ occurs when you flip a figure over a line.

5. Two lines in a plane that never intersect are called _____.

6. A(n) _____ angle is an angle formed at a vertex of a polygon.

7. A(n) _____ is a transformation in which a figure is turned around a fixed point.

8. When two lines intersect, they form two pairs of opposite angles called _____ angles.

9. Two angles are _____ if the sum of their measures is 90°.

10. A(n) _____ is a transformation that enlarges or reduces a figure by a scale factor.

H	T	S	Z	F	R	I	W	Z	D	N	Z	Y	X	I	A	R	T	N	K	I	N	D	P	X
B	N	X	N	K	A	E	W	F	O	W	R	E	N	F	B	B	Y	D	D	T	Q	D	A	V
C	E	O	W	M	W	M	F	I	X	A	L	T	G	V	B	N	O	I	T	A	T	O	R	G
R	C	P	A	H	P	B	T	L	T	J	E	M	J	W	E	L	Y	L	U	D	R	T	A	D
D	A	B	Y	Z	Q	M	R	N	E	R	X	S	V	B	B	R	B	A	I	U	C	W	L	U
X	J	K	L	Y	L	X	E	L	I	C	J	H	Y	M	G	W	T	T	Y	C	U	R	L	S
P	D	A	X	F	O	M	V	O	F	M	T	L	T	C	H	M	G	I	A	P	R	Q	E	B
C	A	Q	A	T	E	B	R	H	P	A	H	I	E	U	P	W	A	O	C	P	C	A	L	K
C	B	A	Q	L	U	E	J	V	Y	U	D	V	O	K	F	Y	P	N	O	A	B	Q	N	R
T	D	Q	P	T	R	A	N	S	L	A	T	I	O	N	U	F	N	Z	H	L	L	L	M	E
R	S	M	Q	I	R	P	N	B	L	P	R	N	S	X	C	R	F	T	Q	M	F	Z	S	M
U	O	Q	X	V	G	G	D	V	D	V	A	W	N	R	R	T	G	P	A	O	H	D	Y	X
C	T	H	C	N	O	I	T	Y	R	A	T	N	E	M	E	L	P	P	U	S	B	G	L	X
A	X	M	F	B	W	J	G	Z	O	K	K	D	I	O	O	M	V	W	G	T	U	O	R	H
Z	C	Y	Z	O	Q	B	Y	C	U	I	G	H	G	A	Z	N	K	K	T	Y	G	P	B	U

Use Your

Use your Foldable to help review the chapter.

Tape here

Tape here

Transformations

Model	Model
Model	**Model**

Tab 1

Tab 2

Got it?

Triangle *ABC* has vertices *A*(2, 1), *B*(2, 4), and *C*(6, 1). Given the vertices of the image, circle the transformation performed.

1. *A*′(4, 2), *B*′(4, 8), *C*′(12, 2)

   translation   reflection   rotation   dilation

2. *A*′(−4, 1), *B*′(−4, 4), *C*′(0, 1)

   translation   reflection   rotation   dilation

3. *A*′(−2, −1), *B*′(−2, −4), *C*′(−6, −1)

   translation   reflection   rotation   dilation

4. *A*′(2, −1), *B*′(2, −4), *C*′(6, −1)

   translation   reflection   rotation   dilation

Problem Solving

1. A decorative floor has the tiling pattern shown. The horizontal segments are all parallel. Lines *AG* and *BH* are parallel and are transversals of the horizontal segments. If the measure of ∠*GDE* = 130°, find the measures of ∠*ADE*, ∠*ADC*, ∠*DAB*, ∠*ABE*, ∠*BEF*, and ∠*BED*. (Lesson 1)

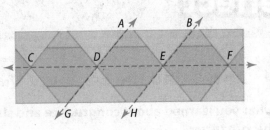

_____

_____

2. The distance between Minneapolis and Omaha is about 378 miles. The table shows the distance between Cincinnati, Ohio, and those two cities. Suppose a triangle was formed by drawing a line between each pair of cities. Classify the triangle by its sides. (Lesson 2) _____

Distance from Cincinnati	
City	Distance (mi)
Minneapolis	692
Omaha	692

3. The cells of a honeycomb are shaped like regular pentagons. (Lesson 3)

a. What is the sum of the interior angle measures of a regular pentagon?

_____

b. What is the measure of an exterior angle of a regular pentagon?

_____

4. Animators can use translations to show movement. Suppose Tayshon draws a ball with a center at (1, 5). Describe the translation if the next screen places the center of the ball at (0, −2). (Lesson 4)

_____

5. Determine whether the quilt pattern shown has rotational symmetry. If so, describe the angle of rotation. (Lesson 5) _____

6. A computer program can dilate photos so they print smaller or larger than the original. (Lesson 7)

a. What scale factor should Ignacio use to dilate a 4 inch by 6 inch photo so it will print as a 5-inch by 7.5-inch photo? _____

b. What scale factor should he use to make a 2-inch by 3-inch print? _____

Reflect

 Answering the Essential Question

Use what you learned about congruence and similarity to complete the graphic organizer.

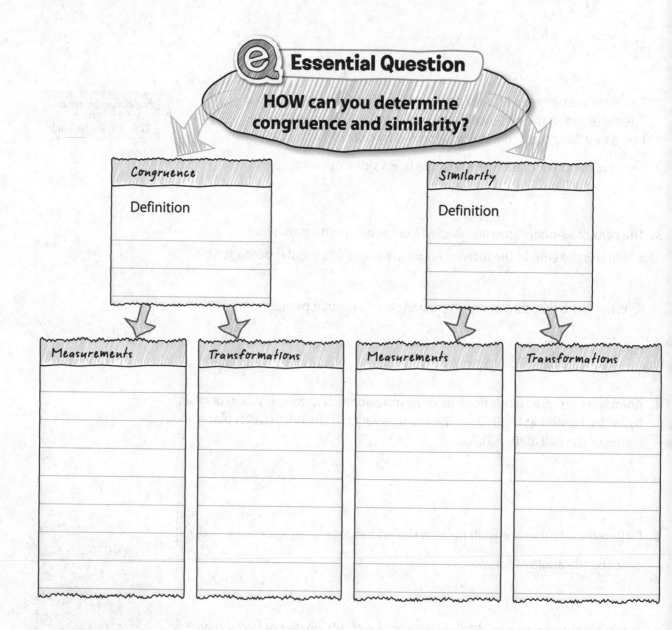

 Essential Question

HOW can you determine congruence and similarity?

Congruence

Definition

Similarity

Definition

Measurements

Transformations

Measurements

Transformations

 Answer the Essential Question. HOW can you determine congruence and similarity?

Chapter 12
Volume and Surface Area

Chapter Preview

 **Vocabulary**

bases	edge	radius
center	face	regular pyramid
circle	lateral area	slant height
circumference	lateral faces	solids
composite figure	pi	sphere
cone	plane	surface area
cross section	polyhedron	vertex
cylinder	prism	volume
diameter	pyramid	

Vocabulary Activity

Use the Glossary to find the definitions of the terms below. Then draw a line to match each term with the correct definition.

1. circle
2. composite figure
3. cross section
4. polyhedron
5. surface area
6. volume

a. A figure that is made up of two or more shapes.

b. A solid with flat surfaces that are polygons.

c. The measure of space occupied by a solid region.

d. The intersection of a solid and a plane.

e. The set of all points in a plane that are the same distance from a given point called the center.

f. The sum of the areas of all the surfaces (faces) of a 3-dimensional figure.

Try the Quick Check below.
Or, take the Online Readiness Quiz.

Check ✓

Common Core Review 6.NS.3, 5.G.3, 5.G.4

Example 1

Find 0.5(3)(6.25). Round to the nearest tenth.

0.5(3)(6.25)

$= [0.5(3)](6.25)$ Order of operations

$= (1.5)(6.25)$ $0.5 \times 3 = 1.5$

$= 9.375$ Multiply.

≈ 9.4 Round to the nearest tenth.

Example 2

Determine whether the figure is a polygon. If it is, classify the polygon.

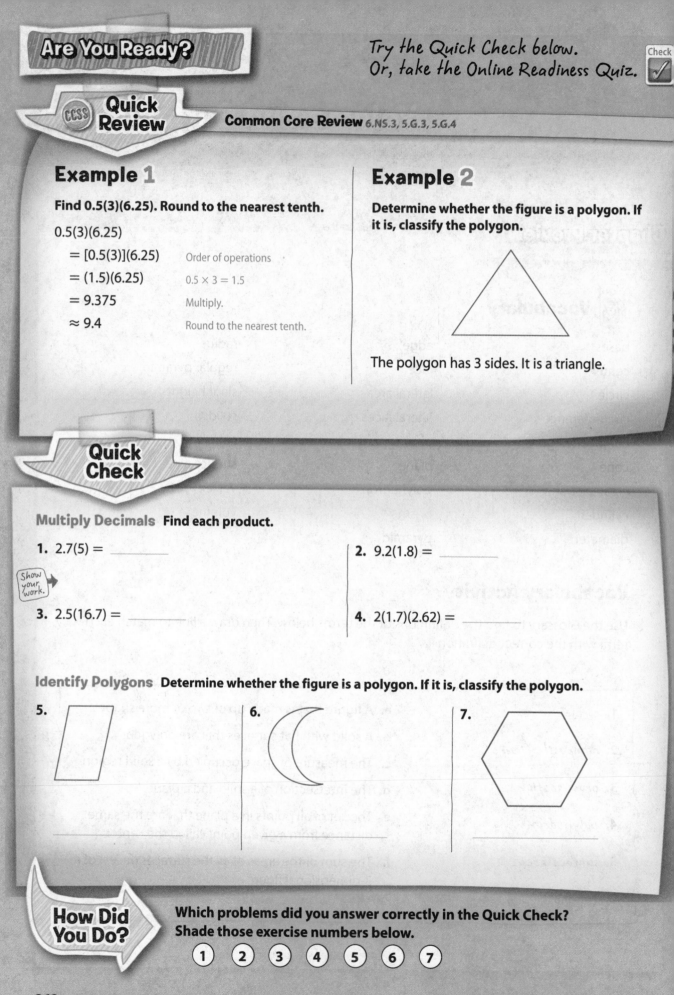

The polygon has 3 sides. It is a triangle.

Quick Check

Multiply Decimals **Find each product.**

1. $2.7(5) = $ _____

2. $9.2(1.8) = $ _____

Show your work. ➡

3. $2.5(16.7) = $ _____

4. $2(1.7)(2.62) = $ _____

Identify Polygons **Determine whether the figure is a polygon. If it is, classify the polygon.**

5.

6.

7.

How Did You Do?

Which problems did you answer correctly in the Quick Check?
Shade those exercise numbers below.

① ② ③ ④ ⑤ ⑥ ⑦

✂ cut on all dashed lines   ⬜ fold on all solid lines   ▨ tape to page 294   **FOLDABLES**®

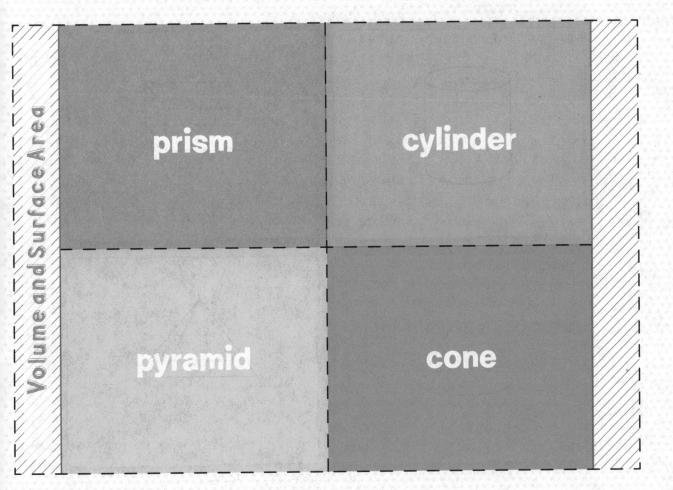

Volume and Surface Area

prism

cylinder

pyramid

cone

FOLDABLES®
Study Organizer

1 Cut out the Foldable above.

2 Place your Foldable on page 294.

3 Use the Foldable throughout this chapter to help you learn about volume and surface area.

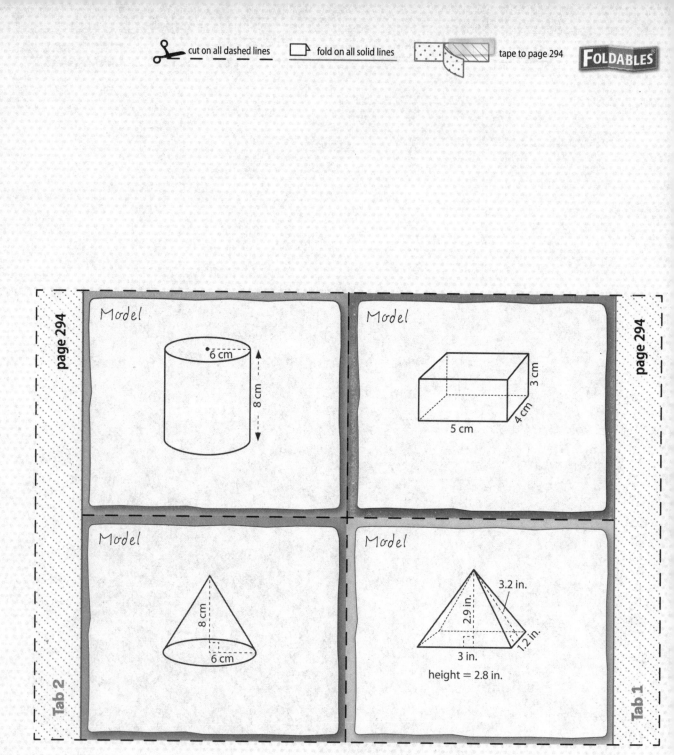

Circles and Circumference

Getting Started

Scan Lesson 12-1 in your textbook. Predict two things you will learn about circles.

- _____
  _____

- _____
  _____

Vocabulary Start-Up

A **circle** is the set of all points in a plane that are the same distance from a given point in the plane, called the **center**. The **circumference** is the distance around the circle. The **diameter** is the distance across a circle through its center. The **radius** is the distance from the center to any point on the circle.

Fill in each box with one of the following terms: *center, circumference, diameter,* **and** *radius.*

Bicycles Bicycle tires, or wheels, vary in size, treads, and smoothness. The type of wheel on a bike depends on how the bike is used—such as for racing, tricks, or touring. All bicycle wheels have hubs and spokes.

Complete the following statements about the construction of a bicycle wheel. Write one of the following terms: *center, circumference, diameter,* **or** *radius.*

1. A spoke of a bicycle tire resembles the _____ of a circle.
2. The spokes of a bicycle are attached to the hub, which is

   the _____ part of the bicycle wheel.

3. The distance covered by one revolution of a bicycle wheel is the

   wheel's _____.

Notes

Circumference of Circles

Cross out the section in the concept circle that does not belong. Then describe the relationship of the three remaining sections.

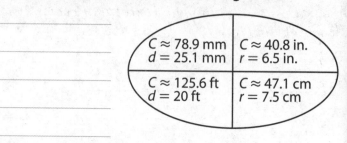

$C \approx 78.9$ mm $d = 25.1$ mm	$C \approx 40.8$ in. $r = 6.5$ in.
$C \approx 125.6$ ft $d = 20$ ft	$C \approx 47.1$ cm $r = 7.5$ cm

Use Circumference to Solve Problems

The radius of a circular fountain is 11.5 feet. Use the four-step plan to find the circumference of the fountain to the nearest tenth.

Understand	Determine what you need to find.	
Plan	Choose a method to solve the problem.	
Solve	Use your plan to solve.	
Check	Check your answer.	

Summary

Write 2–3 sentences to summarize the lesson.

Rate Yourself!

How confident are you about finding circumference? Check the box that applies.

For more help, go online to access a Personal Tutor.

Lesson 12-2
Area of Circles

Getting Started

Scan Lesson 12-2 in your textbook. List two real-world scenarios in which you would find the area of a circle.

- _____

  _____

- _____

  _____

Quick Review
Fill in the blank.
The diameter of a circle
is ____ times the radius.

Real-World Link

Disc Golf Disc golf is one of today's fastest-growing sports. The putting "green" in disc golf is the circular area that is about 33 feet from the disc catcher. The circular area is shown on the grid paper below.

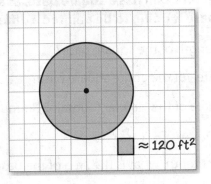

$\approx 120 \text{ ft}^2$

1. Approximately how many squares are shaded? _____

2. If one grid square represents approximately 120 square feet, how many square feet do two grid squares represent? _____

3. If the side of one grid square represents a length of 11 feet, what is the approximate radius of the circle? Explain your reasoning.

   _____

   _____

4. Write a sentence to find the approximate area of a disc golf putting green. Explain your reasoning. _____

   _____

Notes

Area of a Circle

Complete the graphic orgainzer about the area of a circle.

	Area of a Circle
Words	The area A of a circle is equal to _____ _____
Symbols	A = _____
Example	A = _____  5 cm

Use Area of Circles to Solve Problems

A circular water fountain has a square base, as shown in the diagram. The base is going to be painted a different color.

1. What is the length of the square base? _____

2. What is the area of the entire square base? _____

14 in.

3. What is the area of the circular fountain? Round to the nearest tenth. _____

4. To the nearest tenth, what is the area of the base that will need to be painted? _____

Summary

Write 2–3 sentences to summarize the lesson.

_____

_____

_____

_____

_____

_____

_____

_____

Rate Yourself!

How confident are you about finding the area of a circle? Shade the ring on the target.

I'm on target.

I need help.

For more help, go online to access a Personal Tutor.

Area of Composite Figures

Getting Started

Scan Lesson 12-3 in your textbook. List two headings you would use to make an outline of the lesson.

- _____
  _____

- _____
  _____

Quick Review

Write the formula for finding the area of each figure.

circle _____

triangle _____

trapezoid _____

_____

Real-World Link

Pools Above-ground and in-ground swimming pools come in a variety of sizes and shapes. Sometimes the shape is made up of different polygons, as shown in the diagram below.

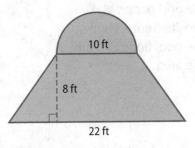

10 ft

8 ft

22 ft

1. Describe the figures that make up the pool. _____
   _____

2. Describe the dimensions that you will use to find the area of each figure.
   _____
   _____
   _____

3. Fill in the blanks to find the area of each figure. Round to the nearest tenth if necessary.

   Area of _____ = _____

   Area of _____ = _____

4. What is the area of the floor of the swimming pool? Round to the nearest tenth. _____

Notes

Area Formulas

Complete the graphic organizer to find the area of the composite figure. Round to the nearest tenth if necessary.

20 m

5 m

•7 m

Find the area of the semicircle.	
Find the area of the rectangle.	
Find the area of the triangle.	
The area of the figure is about _____ .	

Solve Problems Involving Area

Sketch and label your own swimming pool design on the grid paper. Make the shape a composite figure. Then find the area of the pool floor. Explain how you found the area.

_____

_____

_____

Summary

Write 2–3 sentences to summarize the lesson.

_____

_____

_____

_____

_____

_____

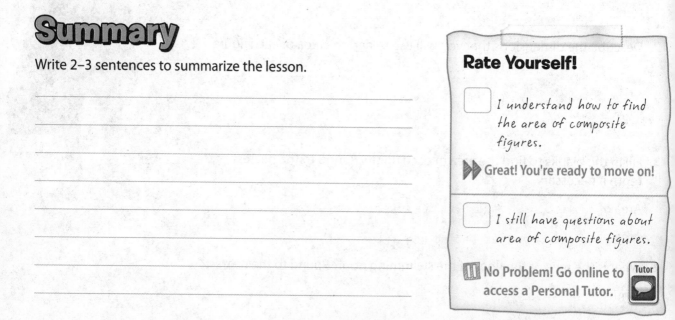

Rate Yourself!

☐ I understand how to find the area of composite figures.

▶▶ Great! You're ready to move on!

☐ I still have questions about area of composite figures.

▌▌ No Problem! Go online to access a Personal Tutor.

Tutor

Three-Dimensional Figures

Getting Started

Scan Lesson 12-4 in your textbook. List two headings you would use to make an outline of the lesson.

- _____
  _____

- _____
  _____

Vocabulary Start-Up

A two-dimensional figure has two dimensions—length and width. A **three-dimensional figure** has three dimensions—length, width, and depth (or height).

Complete the graphic organizer below. For each two-dimensional figure, draw a three-dimensional figure using the given two-dimensional figure as its base.

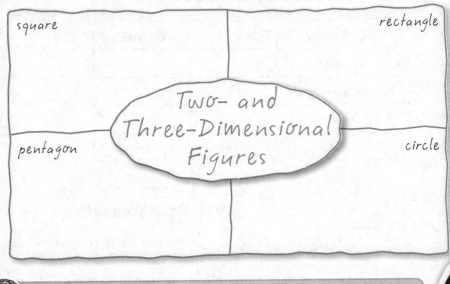

square   rectangle

Two- and Three-Dimensional Figures

pentagon   circle

🌐 Real-World Link

Puzzles Some three-dimensional jigsaw puzzles contain over 3000 foam pieces! These pieces are used to create everything from historical landmarks to everyday objects that are probably found in your home.

Describe the difference between two- and three-dimensional jigsaw puzzles. _____

_____

Notes

Identify Three-Dimensional Figures

Match each term with its definition.

1. edge
2. vertex
3. face

a. where three or more planes intersect in a point
b. a flat surface of a polyhedron
c. where two planes intersect in a line

4. Name the bases, faces, edges, and vertices of the triangular prism.

bases: _____

faces: _____

_____

edges: _____

_____

vertices: _____

Cross Sections

Fill in the graphic organizer on cross sections.

3-D Figure	Type of Slice	Cross Section
cylinder	vertical	
triangular pyramid	horizontal	
cone	angled	
cube	vertical	

Summary

Write 2–3 sentences to summarize the lesson.

_____

_____

_____

_____

_____

_____

_____

Rate Yourself!

Are you ready to move on? Shade the section that applies.

I have a few questions.

I'm ready to move on.

I have a lot of questions.

For more help, go online to access a Personal Tutor. **Tutor**

Volume of Prisms

Getting Started

Scan Lesson 12-5 in your textbook. List two real-world scenarios in which you would find the volume of a prism.

- _____
  _____
- _____
  _____

Quick Review

What is the area of a rectangle with a length of 12 inches and a width of 10 inches?

_____

Real-World Link

Coolers Coolers will keep food cold for days in the hot sun. There are many sizes to choose from. The Bowen family has a cooler like the one shown. Assume that the outside and inside dimensions of the cooler are the same.

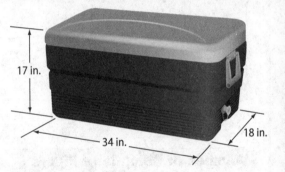

17 in.

34 in.

18 in.

1. What is the area of the bottom of the cooler? _____

2. If you layered the bottom of the cooler with ice cubes that measured 1 inch on all sides, how many ice cubes would you use? _____

3. How many ice cubes would you need to make four layers in the cooler? _____

4. How far up the cooler would those four layers be? _____

5. If you could completely layer the cooler with ice cubes, how many layers would you make? _____

6. How many ice cubes would you need? _____

7. **Volume** is the amount of space inside a three-dimensional figure. The volume of the cooler is ☐ cubic inches. Based on this, what formula could you write to find the volume of a rectangular prism? _____

Notes

Volume of s Prism

Compare finding the volume of a rectangular prism with a triangular prism by completing the graphic organizer.

	Rectangular Prism	Triangular Prism
Formula		
Shape of the Base		
Area of Base		
Model		
Volume		

Volumes of Composite Figures

What real-life figure is a composite figure? Describe how you could find the volume.

_____

_____

Summary

Write 2–3 sentences to summarize the lesson.

_____

_____

_____

_____

_____

_____

_____

Rate Yourself!

Are you ready to move on? Shade the section that applies.

YES   ?   NO

For more help, go online to access a Personal Tutor.

FOLDABLES Time to update your Foldable!

Volume of Cylinders

Getting Started

Scan Lesson 12-6 in your textbook. List two headings you would use to make an outline of the lesson.

- _____
  _____

- _____
  _____

Quick Review

To the nearest tenth, what is the area of a circle with a radius of 4 inches?

_____

Real-World Link

Music Have you ever played music by hitting an ordinary water glass with a spoon? If you want to play a low note, add just a little water to the glass. If you want to play a higher note, add more water.

1. What is the shape of the base of the cylinder? _____

2. What expression can be used to find the area of the base? _____

3. The formula for the volume of a prism is $V = Bh$. Rewrite this formula but

   replace B with the expression from Exercise 2. _____

4. The cylinders below represent glass containers. Based just on how they look, rank the cylinders in order from the least volume to the greatest volume.

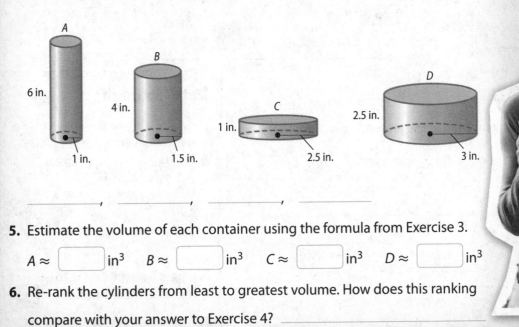

_____ , _____ , _____ , _____

5. Estimate the volume of each container using the formula from Exercise 3.

   $A \approx$ ☐ in³ $B \approx$ ☐ in³ $C \approx$ ☐ in³ $D \approx$ ☐ in³

6. Re-rank the cylinders from least to greatest volume. How does this ranking

   compare with your answer to Exercise 4? _____

Notes

Volume of a Cylinder

Complete the graphic organizer.

How to Find the Volume of a Cylinder	
Words	Multiply the _____ of the cylinder by its _____. The base of a cylinder is a _____.
Formula	
Example	A cylinder has a radius of 4 meters and a height of 7 meters. V = _____ Volume of a cylinder = _____ Substitute. ≈ _____ Simplify.

Volume of Composite Figures

Fill in each blank to complete the steps to find the volume of a composite figure.

Step 1	_____ the figure into _____ figures.
Step 2	Find the _____ of each _____ using the correct _____.
Step 3	Find the _____ of the _____ to find the volume of the _____.

Summary

Write 2–3 sentences to summarize the lesson.

_____

_____

_____

_____

_____

_____

Rate Yourself!

How well do you understand finding the volume of a cylinder? Circle the image that applies.

Clear   Somewhat Clear   Not So Clear

For more help, go online to access a Personal Tutor.

FOLDABLES Time to update your Foldable!

Mid-Chapter Check

Vocabulary Check

1. **CCSS** **Be Precise** Define *three-dimensional figure.* Then give two examples of three-dimensional figures. (Lesson 4) _____

_____

Skills Check and Problem Solving

Find the circumference and area of each circle. Round to the nearest tenth. (Lessons 1 and 2)

2.

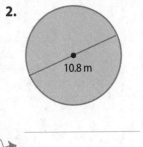

10.8 m

_____

3.

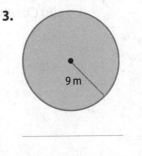

9 m

_____

4.

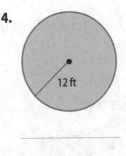

12 ft

_____

Find the volume of each figure. Round to the nearest tenth. (Lessons 5 and 6)

5.

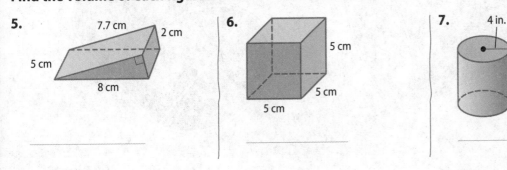

7.7 cm   2 cm

5 cm

8 cm

_____

6.

5 cm

5 cm

5 cm

5 cm

_____

7.   4 in.

8 in.

_____

8. The figure at the right represents a company logo on a T-shirt. What is the area of the logo? (Lesson 3) _____

15 cm

8 cm

9. **Standardized Test Practice** What is the shape resulting from a horizontal cross section of the cylinder in Exercise 7? (Lesson 4)

Ⓐ circle

Ⓑ parabola

Ⓒ rectangle

Ⓓ square

Out of this World Architecture

Use the space laboratories below to solve each problem. Round to the nearest tenth.

1. *Destiny* has one round window that is 20 inches in diameter. What is the circumference of the window? _____

2. What is the area of the window described in Exercise 1? _____

3. If *Columbus* was resting on one of its bases, what shape would result from the vertical cross section of *Columbus*? _____

4. What is the volume of *Destiny*? _____

5. The internal volume of *Columbus*, or the space where the astronauts live and work, is about 34.7 cubic meters less than the total volume. What is the internal volume of *Columbus*? _____

6. *Kibo* is a Japanese laboratory on the International Space Station. It is a cylinder 11.2 meters long with a radius of 2.2 meters. Compare its volume to the volumes of *Destiny* and *Columbus*. _____
_____
_____
_____

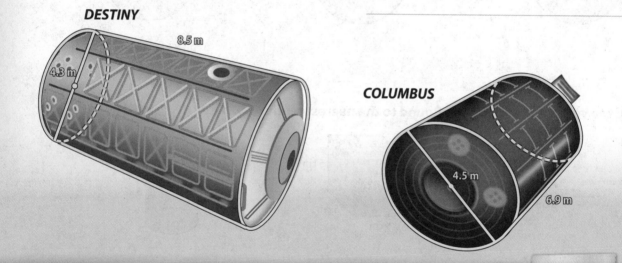

DESTINY
8.5 m
4.3 m

COLUMBUS
4.5 m
6.9 m

Career Project

It's time to update your career portfolio! Use the Internet or another source to research a career as a space architect. Write a paragraph that summarizes your findings.

_____
_____
_____
_____

What subject in school is the most important to you? How would you use that subject in this career?

_____
_____
_____
_____

Volume of Pyramids, Cones, and Spheres

Getting Started

Scan Lesson 12-7 in your textbook. Predict two things you will learn about this lesson.

- _____
  _____
  _____

- _____
  _____
  _____

Quick Review

To the nearest tenth, what is the area of a circle with a radius of 4 inches?

_____

Real-World Link

Basketball Basketballs can be made out of leather, synthetic, or rubber, and they come in different sizes. The diameter of a regulation basketball is about 9.4 inches, which is about half the size of the diameter of the basket.

1. Recall that a *cross section* is the intersection of a three-dimensional figure and a plane. What is the shape of a cross section of the

   basketball shown below? _____

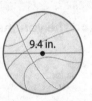

9.4 in.

2. Describe the largest cross section of the basketball.

   _____

3. What is the radius of the largest cross section of the basketball? _____

4. Is this the same as the radius of the basketball? _____

5. What formula is used to find the area of a circle? _____

6. What is the area of the largest cross section of the basketball? Round to the

   nearest tenth. _____

7. The formula to find the volume of a sphere is $V = \frac{4}{3}\pi r^3$. What is the volume of

   a standard basketball? Round to the nearest tenth. _____

Notes

Volume of Pyramids and Cones

Complete the graphic organizer.

Volume of Pyramids and Cones			
	Model	Formula	Example
Pyramid			
Cone			

Volume of a Sphere

Write out each step to find the volume of a sphere with a radius of 3 centimeters.

_____ Formula for the Volume of a Sphere

_____ Replace *r* with [   ].

_____ Simplify.

Summary

Write 2–3 sentences to summarize the lesson.

_____

_____

_____

_____

_____

_____

Rate Yourself!

Are you ready to move on? Shade the section that applies.

YES   ?   NO

For more help, go online to access a Personal Tutor. **Tutor**

FOLDABLES Time to update your Foldable!

Surface Area of Prisms

Getting Started

Scan Lesson 12-8 in your textbook. Write the definitions of lateral faces and surface area.

• lateral faces _____

_____

• surface area _____

_____

Real-World Link

Gift Wrap Oneida is recycling paper bags to make gift wrap paper for a cardboard box like the one shown below. Because of the shape of the box, she decides to trace, cut out, and tape together each of the faces.

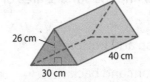

1. The cardboard box has three lateral faces. What shape are the lateral faces?

_____

2. What formula can you use to find the area of a lateral face? _____

3. The cardboard box has two bases. What shape are the bases?

_____

4. What formula can you use to find the area of a base? _____

5. Suppose the base of the cardboard box is an equilateral triangle. Label the cardboard box with the dimensions below. Then fill in the blanks to find how much paper Oneida needs to wrap the box.

length of base: 30 centimeters   height of base: 26 centimeters
length of lateral face: 40 centimeters

surface area = 3 · area of lateral face + 2 · area of base

= 3 · _____ cm² + 2 · _____ cm²

= _____ cm²

Notes

Prisms

1. Complete the graphic organizer on lateral area and surface area.

Lateral Area of Prisms	Surface Area of Prisms
Define It	Define It
Describe How to Find It Multiply _____ _____.	Describe How to Find It Add _____ _____.
Write the Formula L = _____	Write the Formula S = _____

Applying Surface Area

2. Fill in the blanks to find the surface area of the rectangular prism.

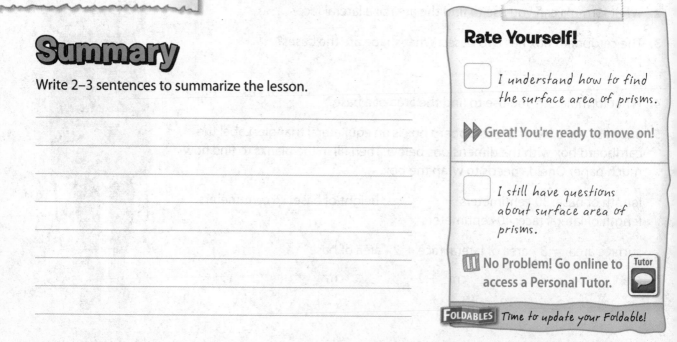

5 cm
5 cm
10 cm

total area of the top and bottom: _____ cm²

total area of the front and back: _____ cm²

total area of the sides: _____ cm²

surface area: _____ cm²

Summary

Write 2–3 sentences to summarize the lesson.

_____

_____

_____

_____

_____

_____

_____

Rate Yourself!

☐ I understand how to find the surface area of prisms.

▶▶ Great! You're ready to move on!

☐ I still have questions about surface area of prisms.

▮▮ No Problem! Go online to access a Personal Tutor. Tutor

FOLDABLES Time to update your Foldable!

Lesson 12-9

Surface Area of Cylinders

Getting Started

Scan Lesson 12-9 in your textbook. List two real-world scenarios in which you would compute the surface area of a cylinder.

- _____
  _____

- _____
  _____

Quick Review

Find the area of a circle with a diameter of 12 centimeters. Round to the nearest tenth.

_____

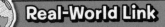

Real-World Link

Food The manufacturer of Super Soups is deciding what size of cylindrical can to make for its new line of healthy soups. In making its decision, the company is considering manufacturing costs and customer feedback.

1. A soup can has two bases. What shape are the bases?

   _____

2. What formula can you use to find the area of one of the

   bases of a soup can? _____

3. A soup can has a curved side. If you remove the label from a soup can, what shape is the label?

   _____

4. What formula can you use to find the area of a soup can label?

   _____

5. Suppose Super Soup decides to manufacture cans that have a radius of 1.5 inches and a height of 4 inches. Label the soup can above and then fill in the blanks to find the amount of metal used in the manufacturing of one soup can. Round your answers for the areas of the circle and curved surface to the nearest tenth.

   surface area = 2 · area of circle + area of curved surface

   $= 2 \cdot$ _____ in^2 + _____ in^2

   $=$ _____ in^2

Notes

Surface Area of Cylinders

1. Draw the net of the cylinder. Label the radius *r*, the height *h*, and the circumference *C* on the net.

2. Explain how drawing the net of a cylinder can help you find the surface area of the cylinder. _____

_____

Compare Surface Areas

3. Which cylinder has the greater surface area? Explain your reasoning.

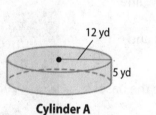

Cylinder A

12 yd

5 yd

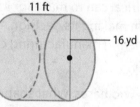

Cylinder B

11 ft

16 yd

_____

_____

_____

Summary

Write 2–3 sentences to summarize the lesson.

_____

_____

_____

_____

_____

_____

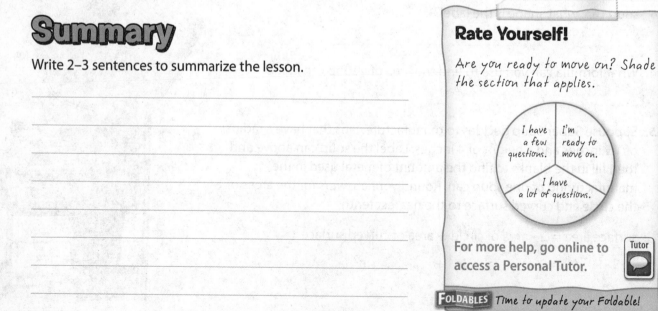

Rate Yourself!

Are you ready to move on? Shade the section that applies.

I have a few questions.

I'm ready to move on.

I have a lot of questions.

For more help, go online to access a Personal Tutor.

FOLDABLES Time to update your Foldable!

Surface Area of Pyramids and Cones

Getting Started

Scan Lesson 12-10 in your textbook. Write the definitions of regular pyramid and slant height.

• regular pyramid _____

_____

• slant height _____

_____

Quick Review
What formula is used to find the area of a circle?

_____

Real-World Link

Ice Cream The first conical-shaped cones were invented by accident at the 1904 World's Fair. An ice cream vendor ran out of paper dishes, so instead, he used rolled waffles from the booth that was next to his. A cone and the net of the same cone are shown below.

Follow these steps to determine the formula for the surface area of a cone.

Step 1 The "wedge" section in the net represents the lateral area of the cone. Divide that section into six equal sections. The first one is done for you.

Step 2 The parallelogram shows the 6 rearranged sections. Write an expression that represents the length of the parallelogram. _____

Step 3 Use the expression from Step 2 to write a formula for the area of the parallelogram, which is the same as the lateral surface area of the cone. _____

Step 4 Write a formula for the total surface area. _____

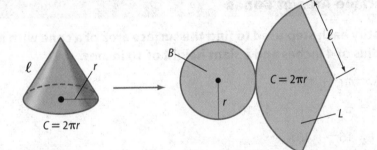

What is the surface area of an ice cream cone that has a diameter of 2 inches and a slant height of 4.5 inches? _____

Notes

Surface Area of Pyramids

Complete the graphic organizer to find the surface area of a pyramid.

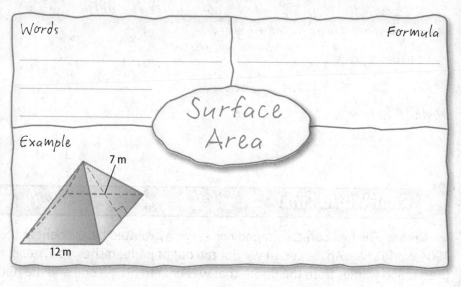

Words

Formula

Surface Area

Example

7 m

12 m

Surface Area of Cones

Justify each step used to find the surface area of a cone with a base radius of 4 inches and a slant height of 10 inches.

$$S.A. = \pi r^2 + \pi r \ell$$

$$= \pi \cdot (4)^2 + \pi \cdot 4 \cdot 10$$

$$= 16\pi + 40\pi$$

$$\approx 175.9$$

Summary

Write 2–3 sentences to summarize the lesson.

Rate Yourself!

How confident are you about finding surface area of pyramids and cones? Check the box that applies.

For more help, go online to access a Personal Tutor.

FOLDABLES *Time to update your Foldable!*

Chapter Review

Complete the crossword puzzle using the vocabulary list at the beginning of the chapter.

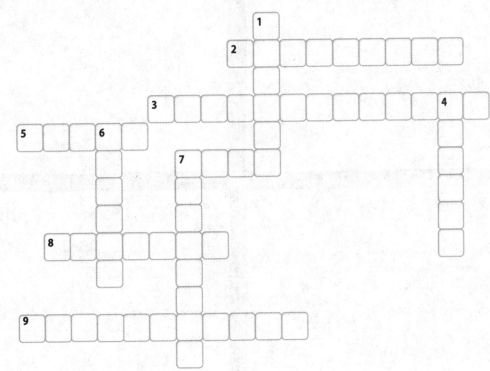

Across

2. a type of figure made up of two or more shapes

3. the distance around a circle

5. a polyhedron that has two parallel, congruent bases in the shape of polygons

7. a three-dimensional figure with one circular base

8. a polyhedron that has a polygon for a base and triangles for sides

9. the area found by adding the areas of all the faces of a 3-dimensional figure

Down

1. the measure of the space occupied by a solid region

4. the set of all points in a plane that are a given distance from the center

6. the set of all points in space that are a given distance from the center

7. a solid that has two parallel, congruent bases connected with a curved side

Use Your FOLDABLES

Use your Foldable to help review the chapter.

Tape here

Tab 1 Volume and Surface Area

Volume

Surface Area

Volume

Surface Area

Volume

Surface Area

Volume

Surface Area

Tape here

Tab 2

Got it?

The problems below may or may not contain an error. If the problem is correct, write a ✓ by the answer. If the problem is not correct, write an X over the answer and correct the problem.

Find the circumference of each circle.

1.
10 cm

The circumference is 62.8 centimeters.

2.
4 in.

The circumference is 25.1 square inches.

Problem Solving

1. Damario is designing a circular fountain to be placed in front of an office building. The fountain will have a decorative stone path around the outside of the fountain as shown.

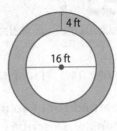

a. What is the circumference of the space occupied by the

fountain and the path to the nearest tenth? (Lesson 1) _____

b. How many square feet of space are needed to build the path?

Round to the nearest tenth. Explain your reasoning. (Lessons 2 and 3) _____

_____

_____

2. Jermal uses the tent shown when he goes camping.

a. Identify the solid. Then name the bases, faces, edges, and vertices. (Lesson 4)

_____

_____

_____

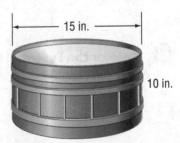

b. Find the volume of the tent. (Lesson 5) _____

c. How much fabric was needed to make the tent? (Lesson 8) _____

3. A giant inflatable balloon used for a parade is spherical in shape. If the balloon has a diameter of 45 feet, how many cubic feet of helium are needed to fill the

balloon? Round to the nearest tenth. (Lesson 7) _____

4. Find the volume and surface area of the drum shown. Round to the

nearest tenth. (Lessons 6 and 9) _____

Reflect

 Answering the Essential Question

Use what you learned about volume and surface area to complete the graphic organizer. Describe how two-dimensional figures are used to find the volume and surface area of prisms, pyramids, cylinders, and cones.

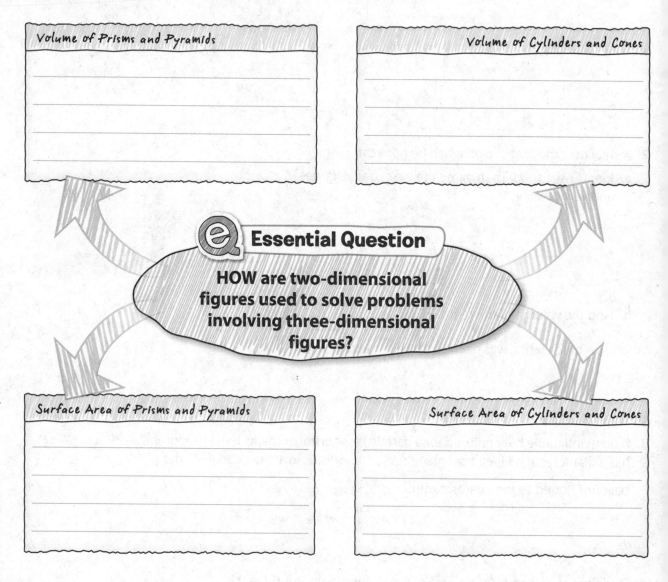

Volume of Prisms and Pyramids

Volume of Cylinders and Cones

Essential Question

HOW are two-dimensional figures used to solve problems involving three-dimensional figures?

Surface Area of Prisms and Pyramids

Surface Area of Cylinders and Cones

 Answer the Essential Question. HOW are two-dimensional figures used to solve problems involving three-dimensional figures?

_____

Mathematics Reference Sheet

mulas

meter	square	$P = 4s$
	rectangle	$P = 2\ell + 2w$ or $P = 2(\ell + w)$
mference	circle	$C = 2\pi r$ or $C = \pi d$
	square	$A = s^2$
	rectangle	$A = \ell w$
	parallelogram	$A = bh$
	triangle	$A = \frac{1}{2}bh$
	trapezoid	$A = \frac{1}{2}h(b_1 + b_2)$
	circle	$A = \pi r^2$
	cube	$V = s^3$
	rectangular prism	$V = Bh$ or $V = \ell wh$
	triangular prism	$V = Bh$
me	cylinder	$V = Bh$ or $V = \pi r^2 h$
	pyramid	$V = \frac{1}{3}Bh$
	cone	$V = \frac{1}{3}Bh$ or $V = \frac{1}{3}\pi r^2 h$
	sphere	$V = \frac{4}{3}\pi r^3$
	cube	$V = 6s^2$
	prism	**Lateral Area:** $L = Ph$ **Surface Area:** $S = L + 2B$ or $S = Ph + 2B$
ace Area	cylinder	**Lateral Area:** $L = 2\pi rh$ **Surface Area:** $S = L + 2B$ or $S = 2\pi rh + 2\pi r^2$
	pyramid	**Lateral Area:** $L = \frac{1}{2}P\ell$ **Surface Area:** $S = L + B$ or $S = \frac{1}{2}P\ell + B$
	cone	**Lateral Area:** $L = \pi r\ell$ **Surface Area:** $S = L + B$ or $S = \pi r\ell + \pi r^2$
perature	Fahrenheit to Celsius	$C = \frac{5}{9}(F - 32)$
	Celsius to Fahrenheit	$F = \frac{9}{5}C + 32$
e	line	$m = \dfrac{\text{rise}}{\text{run}}$ or $m = \dfrac{\text{change in } y}{\text{change in } x}$ or $m = \dfrac{y_2 - y_1}{x_2 - x_1}$

Measures

Length	1 kilometer (km) = 1000 meters (m) 1 meter = 100 centimeters (cm) 1 centimeter = 10 millimeters (mm)	1 foot (ft) = 12 inches (in.) 1 yard (yd) = 3 feet or 36 inches 1 mile (mi) = 1760 yards or 5280 feet
Volume and Capacity	1 kiloliter (kL) = 1000 liters (L) 1 liter = 1000 milliliters (mL)	1 cup (c) = 8 fluid ounces (fl oz) 1 pint (pt) = 2 cups 1 quart (qt) = 2 pints 1 gallon (gal) = 4 quarts
Weight and Mass	1 kilogram (kg) = 1000 grams (g) 1 gram = 1000 milligrams (mg) 1 metric ton (t) = 1000 kilograms	1 pound (lb) = 16 ounces (oz) 1 ton (T) = 2000 pounds
Time	1 minute (min) = 60 seconds (s) 1 hour (h) = 60 minutes 1 day (d) = 24 hours	1 week (wk) = 7 days 1 year (yr) = 12 months (mo) or 52 weeks or 365 day 1 leap year = 366 days
Metric to Customary	1 meter ≈ 39.37 inches 1 kilometer ≈ 0.62 mile 1 centimeter ≈ 0.39 inch	1 kilogram ≈ 2.2 pounds 1 gram ≈ 0.035 ounce 1 liter ≈ 1.057 quarts

Symbols

Number and Operations

$\left.\begin{array}{l} a \cdot b \\ a \times b \\ ab \\ a(b) \end{array}\right\}$ a times b	$\neq$	is not equal to	$\approx$	is approximately equal t
	$>$	is greater than	$\pm$	plus or minus
	$<$	is less than	%	percent
	$\geq$	is greater than or equal to	$a:b$	the ratio of a to b, or $\frac{a}{b}$
	$\leq$	is less than or equal to	$0.7\overline{5}$	repeating decimal 0.755

Algebra and Functions

$-a$	opposite or additive inverse of a	a^{-n} $\frac{1}{a^n}$		$\sqrt{x}$	principal (positive) squa root of x
a^n	a to the nth power	$|x|$	absolute value of x	$f(x)$	function, f of x

Geometry and Measurement

$\cong$	is congruent to	AB	the length of $\overline{AB}$	$\triangle ABC$	triangle ABC
$\sim$	is similar to	∟	right angle	O	origin
°	degree(s)	$\|$	is parallel to	π	pi $\approx \left(3.14 \text{ or } \frac{22}{7}\right)$
$\overleftrightarrow{AB}$	line AB	$\perp$	is perpendicular to	(a, b)	ordered pair with x-coor a and y-coordinate b
$\overrightarrow{AB}$	ray AB	$\angle A$	angle A		
$\overline{AB}$	line segment AB	$m\angle A$	measure of angle A		

Probability and Statistics

$P(A)$	probability of event A